P9-CQX-068

The Ocean World of Jacques Cousteau

Challenges of the Sea

The Ocean World of Jacques Cousteau

Volume 18

Challenges of the Sea

THE DANBURY PRESS

*With only a mouth as his defense, a **jawfish** keeps a wary watch on all activity around his home. In meeting the challenges of the sea both man and marine life have created innovative solutions to their problems, in some cases involving cooperation and in others antagonism.*

The Danbury Press
A Division of Grolier Enterprises Inc.

Publisher: Robert B. Clarke

Production Supervision: William Frampton

Published by Harry N. Abrams, Inc.

Published exclusively in Canada by
Prentice-Hall of Canada, Ltd.

Revised edition—1975

Project Director: Steven Schepp

Managing Editor: Richard C. Murphy

Assistant Managing Editor: Christine Names
Senior Editors: Ellen Boughn
Ralph Slayton
Scientific Consultant: Robert Schreiber
Editorial Assistant: Joanne Cozzi

Art Director and Designer: Gail Ash

Assistants to the Art Director: Martina Franz
Leonard S. Levine
Illustrations Editor: Howard Koslow

Creative Consultant: Milton Charles

Printed in the United States of America

12345678998765

LIBRARY OF CONGRESS CATALOGING
IN PUBLICATION DATA

Cousteau, Jacques Yves.
Challenges of the sea.

(His The ocean world of Jacques Cousteau; v. 18)
1. Oceanographic research. 2. Marine resources conservation. I. Title.
GC57.C68 301.31 74-23066
ISBN 0-8109-0592-2

Contents

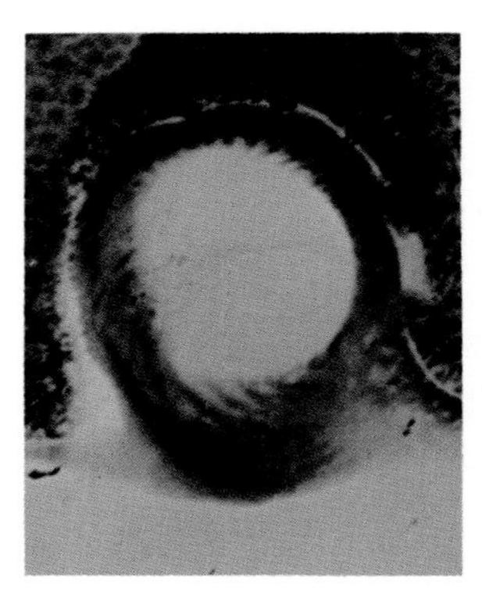

Asian tribes onto the continent of North America. Today those ancient pathways are shallow seas. 62

The sea has always been a challenge to the men and women who ventured out on it. Many times the sea has defeated them. The bottom is littered with the remains of countless wrecks. Many times, with great courage and a strong will to live, MEN AGAINST THE SEA (Chapter VI) have survived challenges that appeared insurmountable. 72

Beneath the sea lie hidden many of the answers to our questions about man's past. They are in the ships that have sunk and in the cargoes they carried. They are in harbors or entire cities that either sank or were drowned by rising seas. A few of these sites have been successfully excavated by the archaeologists DIVING TO THE PAST (Chapter VII). Many others await their investigation. An even greater number of them still wait to be discovered. 102

On the walls of a temple of Ramses III at Madinet Habu near Thebes can be seen the earliest surviving representation of a battle at sea. It is a scene from the Pharaoh's victory over "the people from the sea," "the northerners of the isles," who invaded Egypt with the help of a large fleet in about 1190 B.C. It is the earliest surviving illustration that we have of THE SEA THAT DIVIDES MEN (Chapter VIII). Since that time, mastery of the seas has changed hands many times and so have the ways of naval warfare. 114

For tens of thousands of years, boats were no more than fallen logs paddled by hand. Then one day someone somewhere thought of tying two logs together, and he made the first raft. In another part of the world a man discovered that an armful of reeds was enough to keep him afloat. From the beginning, he used the waterways to get from one place to another and to other men, on THE SEA THAT UNITES MEN (Chapter IX). 126

There is little that we can do to reap the potential harvest of the sea or minimize its destructive forces until we understand it far better than we do now. Of all the CHALLENGES OF THE FUTURE (Chapter X), the greatest is to mobilize our resources and our talents in order that we might better understand the intricate life of the oceans. 132

The ocean—its might, its potential, its mysteries—provides man with some of the greatest challenges he will ever face in his ENDEAVOR TO LIVE.

Introduction: Seeking a More Abundant Life

Ever since the first living cell divided in two, all the creatures of ocean and earth thrived—thanks to, and at the same time in spite of, their environment. Nature was lavish with both opportunities and obstacles, and to live meant simply to take full advantage of all opportunities and to cope with or, better yet, overcome the obstacles. Survival was a permanent challenge for both the individual and the species. For any individual, the bare essentials were to grow and to last as long as possible and, for any species, to ensure reproduction. But beyond survival, all living things strive for a better life. The awesome migrations of some birds, fish, or mammals are the result of millions of daring attempts by their ancestors to seek for the best, the most secure, the most comfortable conditions the planet could offer.

For 3 to 3½ billion years, the physical, instinctive, and preintelligent forms of life have met the challenges of the surrounding world; those that were successful have achieved a certain degree of security and access to as large a span as they could conquer. This pattern of existence, at the same time generous and rigorous, was born in the sea long before it was extended on land, and it is not surprising that it is in the sea that the most formidable challenges of nature have been successfully overcome.

Changes in the conditions of habitats have always presented an almost infinite variety of conveniences and of sudden threats. Most probably the diversity of natural challenges triggered the incredible flexibility of the evolutionary processes and helped generate the hundreds of thousands of different plants and animals. Before we extend our remarks to human beings, we have to realize the built-in fecundity of hardships. In a handful of soil, there are billions of microbes, constantly struggling against adversity—too much rain, heat, cold, or drought. Given the innumerable ways in which life is challenged, it is amazing to think of the number of creatures that win the battle for life. The whole earth and all its creatures are intimately involved in a dynamic adventure in which every happening has an effect, however remote, on everything else; and, of course, man is included in this universal dynamic interrelationship. It is overwhelming—or consoling—to realize that we are all dependent on one another in the struggle to survive.

The advent of civilized man modified the pace and the nature of his own evolution. It is highly probable that the success of *Homo sapiens* is not the direct result of the evolution of his brain but, as we have discussed in the volume *Instinct and Intelligence,* is due to the simultaneous development of brain, hands, articulated voice, and longevity. These combined abilities gave birth to civilization, which is the storehouse of accumulated experience and the access to it. Men could not develop as human creatures, independently from their cultural environment. Modern civilization created the means to fulfill easily the basic needs of all men: food, shelter, clothing, health, and education. Although these essentials are not shared by all, their universal availability remains a major goal of civilized men.

Having—at least theoretically—eliminated the natural challenges of physical life, we have turned to spiritual or intellectual challenges in order to increase the scope and the quality of our life. Reaching for the moon or for the bottom of the ocean is a form of the universal quest for a wider life—the same drive that started the ancestors of the migrating birds. Why would the eels of Europe and of North America travel all the way to the Sargasso Sea to lay their eggs? Why would the salmon hurdle rapids to get at its freshwater spawning ground? Why would a sane man choose to go around the world in a rowboat? Most of the challenges we face today are man-made: overpopulation, waste of resources, and destruction of our environment. Our deep motivation is to overcome our problems, even if they originate in ourselves. It is the search, the fight, not the achievement or the victory that provides us with the closest thing to happiness. We cannot do without challenges. If there were none, the world would become meaningless—it would simply obey the second law of thermodynamics and slowly grind to a halt.

Jacques-Yves Cousteau

Chapter I. Aeons Before Man

Aeons before man ventured out on the open sea, thousands of other creatures were striking out over the ocean on journeys thousands of miles long. Some early sailors were aware of the migratory flights of birds and depended upon these creatures as their only method of navigating the seas. Much later, when man finally learned the basic principles of navigation, he still had to struggle for centuries to untangle the patterns of the stars—a process that birds, we know now, understood instinctively.

All animals move. Some, like a few pelagic sharks, continuously roam the seas in an irregular, nonrepeating search for food. Like hoboes, they are nomads. In human society political and social conditions sometimes cause great numbers of individuals to permanently leave their homeland—to emigrate. Catastrophic events move animals in the same way. Volcanic eruptions send Hawaiian reef fish away from home to seek a more hospitable location. None of these movements should be confused with the periodic, repeating travels, called migration, that whales, eels, shad, salmon, and arctic terns, among others, make to ancestral birthplaces or feeding grounds.

"Recently animal migration has become more than a biological curiosity. Engineers now look to animal navigation for clues to help design submarine guidance systems."

It is remarkable that so many different groups of animals have developed migration patterns and the ability to navigate. These innate behavioral traits are seen in animals as distantly related as the butterfly and the eel. The same behavior has arisen independently in many animals, and a number of navigational techniques serve various organisms equally well.

By whatever means an animal manages to find its destination, we can be sure that over many hundreds of generations it became imperative to the survival of the species to engage in periodic movements. The animals which chose to stay put perished and their stubborn refusal to leave home was not perpetuated in future generations.

Biologists agree that three basic stimuli exist for migration: gametic (reproductive), alimental (dietary), and climatic. Organisms that move to distant locations to reproduce, like sea turtles and American and European eels, are responding to reproductive needs. When whales and sea turtles return from reproductive migrations, they are engaging in alimental migration: they return to feeding grounds. Climatic migration is often related to alimental. Migratory fish, for example, sometimes follow food sources that are regulated by climate. Usually migratory behavior is related to subtle combinations of all three stimuli.

Animal migration has recently become more than a biological curiosity. A new science has developed—bionics, the adaptation of animal mechanisms to human systems. Engineers now look to biological navigational systems to help them devise more sensitive means of guiding submarines through undersea canyons and aircraft to precise locations.

A secluded Costa Rican beach is the scene for ***mass nesting of Pacific ridley turtles,*** *a species of sea turtle in grave danger due to a high demand, worldwide, for its shell, skin, and eggs.*

No Place Like Home

Each year in streams as scattered as those of Japan, Siberia, Scotland, and North America thousands of salmon propel themselves against prevailing currents to fight their way back from the sea to mate in the stream where they were born. Their determination is so great that no obstacle can induce them to give up the battle and return to sea.

Salmon homing instincts were first linked to the sense of smell in experiments which proved that the fish are capable of distinguishing the odors of different streams. Definitive evidence that pointed to salmon navigation by the olfactory sense was assembled in a final experiment that took place in an actual stream bed.

Salmon were removed from their home base and placed downstream at a fork in the Issaquah River. Before they were released to make their upstream migration, the nostrils of half of the fish were plugged with cotton. Most of the control fish easily found their way home, but the smell-blinded salmon picked the wrong stream as often as not.

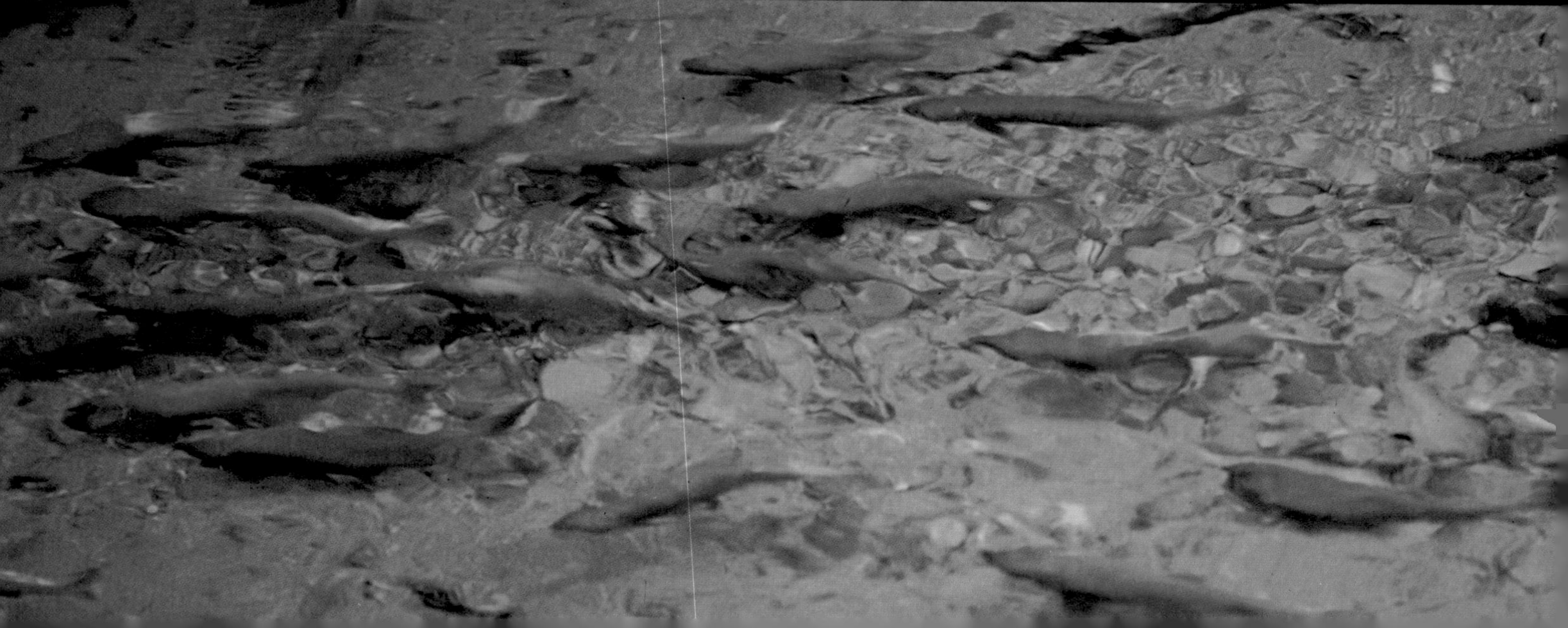

Pacific salmon *(right and below) meet one of the greatest challenges of the sea by performing an amazing migration complicated by physical suffering. After eggs are laid and fertilized, the spent adults will rapidly age and then die of exhaustion.*

Mating turtles *arrive on their traditional nesting grounds, after having navigated thousands of miles of open sea from their normal habitats.*

Mysterious Pathfinders

There are seasons in the sea just as on land. Food supplies and water temperatures, especially in temperate and arctic regions, vary during the year's cycle. Some fish respond to these changes as regularly as do migrating birds. The shad, largest fish in the herring family, times its annual spawning runs into freshwater streams to seasonal changes in water temperature. The more northern shad migrate out of the sea later in the spring than those that travel into rivers in the southern part of the United States. Shad only show up in freshwater streams when the water temperature there falls in a range between 13° and 18° C. (55.4° to 64.4° F.). There is evidence that shad populations also follow a north-south migration in response to seasonal changes in temperature. Since all the members of the breeding population in one area answer the same temperature-regulated spawning call, most arrive at the breeding grounds at the same time—an obvious help in ensuring the continuance of generations.

Green sea turtles appear on island breeding grounds at certain times each year in the Caribbean and South Atlantic. Experimental evidence indicates that seasonal variances in water temperature are not the stimulus that triggers the turtles' migratory urge. Dr. Archie Carr of the University of Miami is responsible for uncovering what is known about the life of the green turtle. Fifteen years ago he and his students set out to tag turtles on Ascension Island, the main rookery in the South Atlantic. Their efforts were to shed light on the life and migratory habits of these elusive animals.

Both male and female turtles push their heavy bodies 1500 miles across the Atlantic to Ascension once every three or four years throughout their lifetime. The rites of mating take place in the surf off the island. Even though the males have made an exhausting migration and often engage in violent mating confrontations with other turtles, they do not come on shore to feed or rest. They await the females who come off the island to breed after laying their eggs.

As soon as the infant turtles hatch, they head in the direction of the sea, even though they may not be able to see it from their nest. Carr and his associates have made some progress in understanding the seaward orientation of the hatchlings. In a tank located between a bay and the ocean on the island of Bimini, the turtles piled up in the area of the tank closest to the bay—the nearest body of water. When the sun's reflection off both the bay and the ocean was obscured, they swam aimlessly. They can seemingly determine the location of water by the brightness of the sky as related to surface reflection.

Navigational methods used by adult turtles to find their tiny island in the midst of thousands of miles of open sea remains a mystery. Some researchers have concluded that turtles have an olfactory sense that guides them to the rookery. Indeed females coming ashore to nest are often seen poking their noses into the sand as if to sniff out the territory. But it does not seem likely that the island could be detected from very far distances by olfactory cues alone.

The current theory is that the turtles travel up and down the coast of Brazil until they locate (by smell and visual clues) the place where they made their first landfalls in youth. Then a compass sense, combined with an ability to orient from the position of the sun as it moves across the sky, brings them close to Ascension; then olfactory and visual messages guide them in.

It is the belief of some scientists that eels, who spawn in the Sargasso Sea southeast of Bermuda, as well as some salmon and sharks, may navigate by use of the weak electric fields generated in the ocean by the movement of currents through the earth's magnetic field. Laboratory experiments indicate that eels are sensitive to currents that are within the voltage generated by the movement of the major ocean currents. Not only can they sense the direction of the current, they may also be able to register, electronically, an upstream or downstream flow.

__A green turtle__ emerges from its shell. Reflections from the nearest body of water will orient the hatchling on a direct line to the sea.

***Arctic terns** (above and below) are champion migrators. Their round trip from Arctic Circle to antarctic wintering grounds may cover 22,000 miles.*

Solar and Celestial Clues

The record for long distance migration belongs to the arctic tern. Annual round-trip distances from nesting sites in the arctic to winter grounds in the antarctic can be as much as 22,000 miles. Northern nesting areas are circumpolar and extend as far south as Cape Cod. Breeding takes place after colonies come together in early May. In a month the eggs hatch. After only a few weeks' diet of fish, the nestlings join their parents in the long haul to the south.

Birds living in northern Europe fly down past the west coast of Africa to Antarctica. North American terns similarly congregate near the antarctic ice pack after flying down to the west of the American continent. The tern is engaged in migratory travels a good two-thirds of the year and is clearly guided by a highly sophisticated navigational system. Since many fly out over the open sea, where even their high-altitude vantage point gives them little information about their location, their navigational tools are of great interest. The first clues to avian navigational methods were found in 1949 by Gustav Kramer, a German ornithologist. He observed that during the day in the spring caged starlings faced towards the northeast, the direction that starlings migrate in the wild. Kramer's birds could see only the sky. Apparently the sun was guiding them.

Kramer manipulated the angle of the sun's rays with mirrors to imitate seasonal changes. The birds shifted their position. But this exciting discovery did not explain how night-flying migrants find their way without cueing on the sun.

An incredible explanation for nocturnal navigation was given in 1955. Franz and Eleanore Sauer observed that warblers, given an unobscured view of the nighttime sky, oriented in the direction of their regular migratory paths. When the sky was heavily overcast, the birds ceased to orient. To prove that the birds used the stars as guiding lights, a process that eluded man for so long, the Sauers took their birds into a planetarium where the features of the sky at any season in any locality could be projected. A springtime sky was presented to a warbler; he faced northeast. A bird that had spent his life in a cage instinctively oriented toward his species' migratory route.

The ability that displaced birds have to find home was also explained. When the warbler was given a sky that corresponded to a location in Siberia, where he had most probably never been, he became confused for a few minutes and then flapped in a direction that would have brought him to Germany—the starting point of his migration. The sky is always changing, and the fact that the birds recognize the change and relate it to their present locality is remarkable.

Migrating geese *(above) rest in a refuge.* ***Snow geese*** *(below) breed in the arctic tundra and, with their young, travel south to winter.*

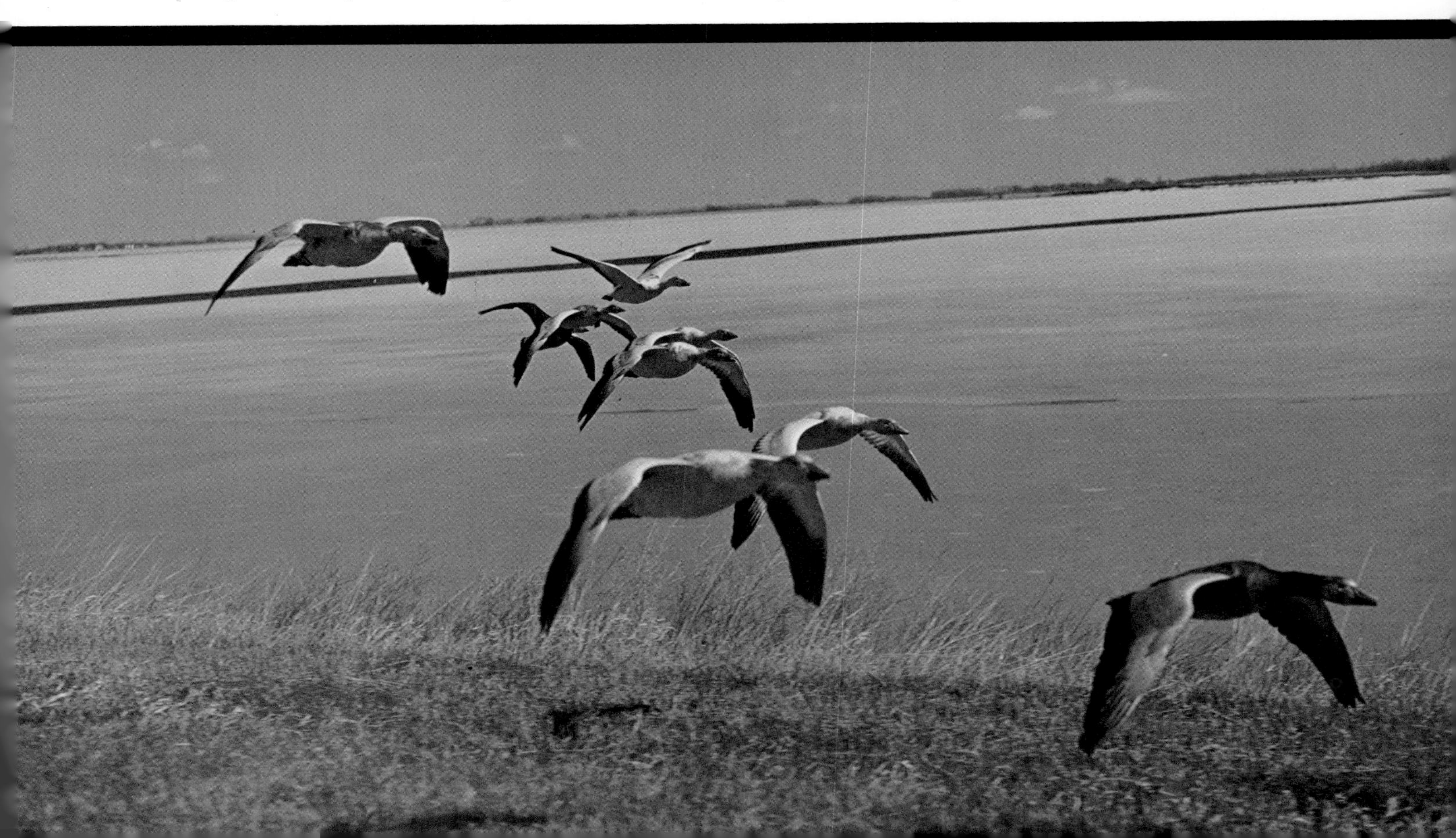

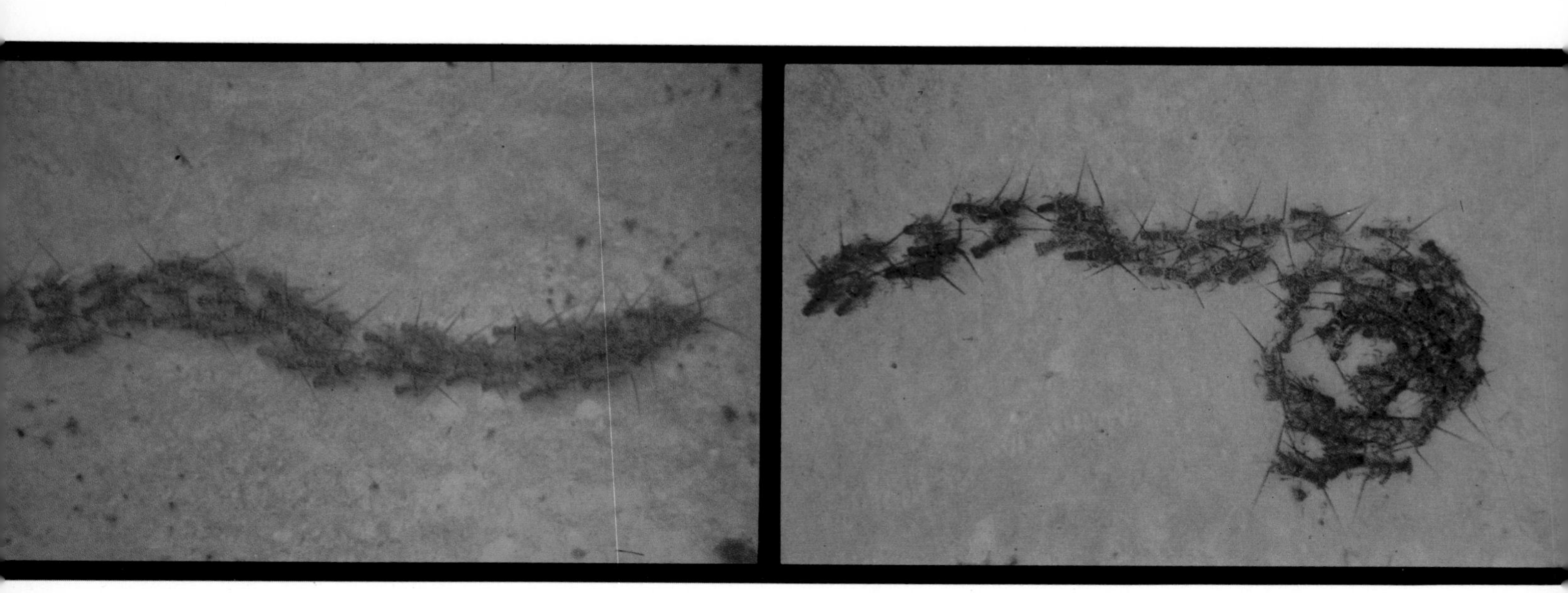

Marching Lobsters

One of the most organized migratory behaviors is that of the spiny lobster in the Caribbean. The migrating crustaceans form single-file lines, some measuring over a quarter-mile long and containing over 1000 lobsters. The head lobster, who appears to gain his position because he is the most eager migrator, is kept in contact with each succeeding member of the line by touch. The head and thorax of one lobster extend over the abdomen of the next in line. The tips of the front legs are hooked around the tail of the lobster in front. When more than one line forms, all lines set off together in parallel formation as if marching to the same drummer.

A diving-scientist, William Herrkind, has extensively examined the lobster phenomenon and has collected reports of other observers. In one of his experiments a number of lobsters were removed from the line and placed in a seawater pool. The group immediately formed a line of march clockwise around the circular tank. They kept walking, day and night, for nearly five weeks. Only small amounts of food were taken and the normally reclusive animals ignored shelters that they would have ordinarily welcomed.

Lobsters taken from the migratory line and released at sea two miles away at a depth of 1500 feet took only two weeks to get back to the point of capture. The lobsters' navigational system isn't related to vision. Blinded lobsters orient just as well as normal ones. When the antennae or walking legs that the animals use to establish contact with others are taped so as to be useless, the lobster substitutes remaining appendages.

The question that puzzles lobster-parade watchers the most is why do the lobsters, who usually shun social behavior, get together for their periodic walks. One theory is that overpopulation in an area acts as a stimulus to movement. A flaw in this theory is that all the lobsters in a certain area move out together, taking their population problem with them.

It has been suggested by biologist Robert Schroeder that lobsters migrate for nutritional reasons. A steady diet of fish drives the lobsters to a mass exodus, while those fed molluscs are content to stay where they are. Environmental changes associated with storms appear to be the only common denominator of most lobster marches. Often after a few days of turbulent weather, the lobsters are on the move.

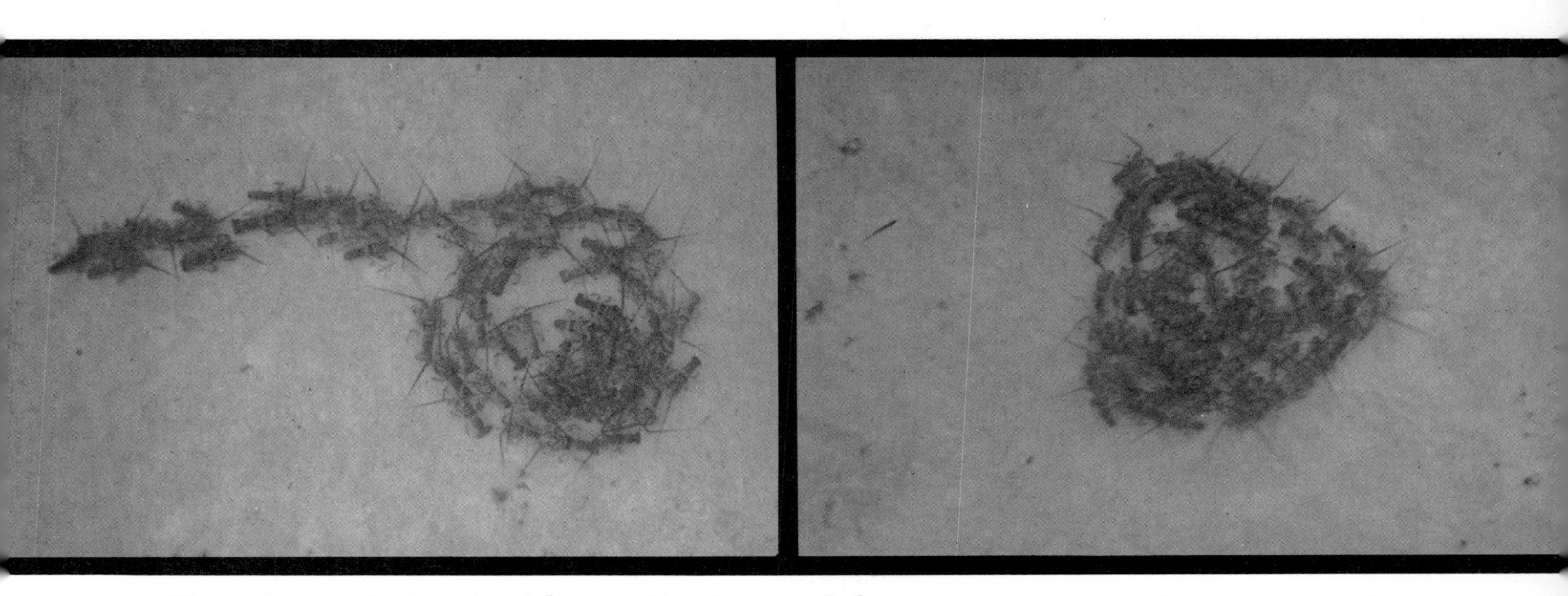

*When interrupted, **migrating lobsters** follow their leader into a tight pod formation (above, left to right) that affords them line protection.*

***Lobsters** taken from a march across the sea floor re-form in a laboratory tank (below). One such group maintained the march continuously for five weeks.*

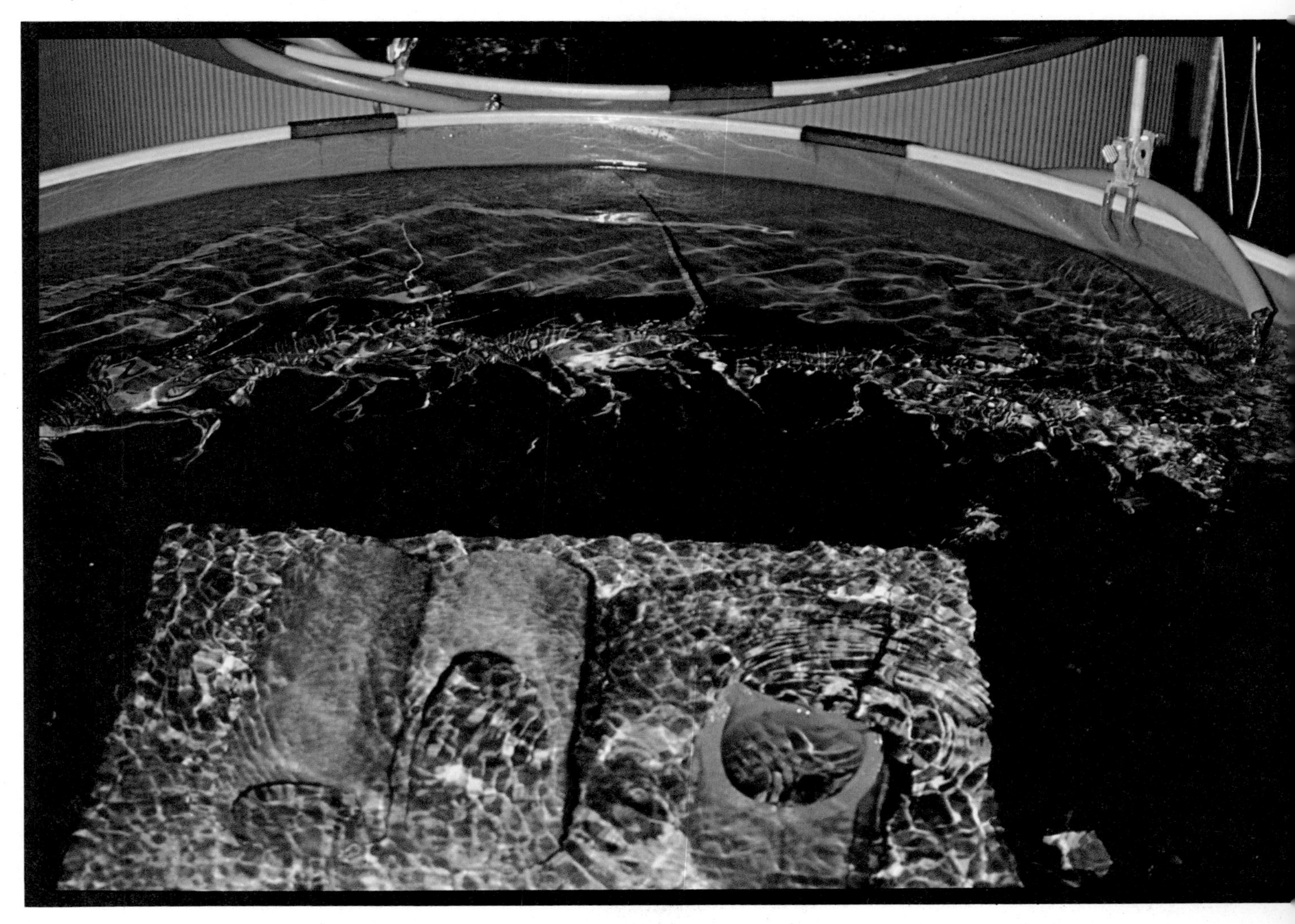

Living Clocks

Navigation by sun and stars requires a built-in time sense. The sun moves across the sky each day. If a migrating animal simply followed the solar path, its migratory route would double back on itself beginning at noon. The animal must be able to adjust the angle of its direction relative to the sun's position according to a daily rhythm. It must know the local time in order to make navigational corrections for the local angle of the sun. Celestial navigation must also be based on time: the stars change their positions with the hours and the seasons.

The lives of many animals are regulated by internal clocks. The urge to migrate at a certain time of the year is apparently a response to length of day. Called photoperiodism, it was demonstrated by inducing migration out

*The color of **fiddler crabs** changes with the fluctuating tides. Crabs in the laboratory exhibit the same rhythms as these (top and below) in the wild.*

of season in birds that were subjected to artificially lengthened periods of daylight.

Migratory journeys are not the only seasonal changes in the living world that are responses to a time sense. The reproductive behavior of an unusual worm called the palolo worm is timed by the moon. Each year during the first days of the full moon in October and again in November vast swarms of the worms surface above their Pacific reef habitats at dawn. Only the hindpart of the adult worm rises to reproduce. This portion disengages from the head, which stays alive in the reef, and comes to the surface to shed eggs and sperm and then disintegrates.

The cycles of the moon control the lives of all tidal organisms since the tides themselves are regulated by lunar cycles. The reproductive organs of sea urchins and oysters enlarge and mature in response to lunar periods. The fiddler crabs' daily feeding habits are determined by the tides as they feed when the tide is out. The very color of the animals is timed to a 24-hour clock. Their bodies darken in the morning and lighten in the evening. (Such 24-hour cycles are called **circadian rhythms.**) Scientists were astounded to discover that levels of activity, as well as the color change, of fiddler crabs which were kept in the laboratory far from their home maintained the tidal times. Compensations were even made for the fact that the tide is 50.5 minutes later every day.

A sea lily (crinoid) runs on an 18-year cycle. Even single-celled plants and animals seem to know what time it is. Both the nucleus and the cytoplasmic stuffing of the single-celled algae, *Acetabularia,* possess a clock.

A remarkable aspect of biological clocks is their temperature independence. It seems only logical that whatever the clock is, it must be controlled by or be part of a chemical reaction as are the rest of our physiological and even mental activities. The catch is that all known chemical reactions are speeded by an increase in temperature and slowed by a decrease in temperature. Such fluctuations would render a clock useless, so obviously biological clocks are not affected by temperature changes. The question remains: How can this biological clock flout a basic principle of chemistry?

Acetabularia *(above) is a large unicellular algae. Both the cytoplasm and the nucleus have biological clocks that regulate photosynthetic cycles.*

Since the time sense of many organisms is maintained when they are removed from their natural environment and from solar, lunar, and tidal clues, some scientists believe that the electromagnetic fields of the earth and the influences of the celestial bodies may be the mainspring of living clocks.

Chapter II. Social Challenges

All animals, including human beings, are social creatures, acting and reacting to the behavior of others of their own species. Social life has both advantages and limitations, and at times it also requires sacrifices. It leads quite naturally to friction between individuals and groups. To live in a society is a challenge in itself.

There are rules for all societies, some written, some only vaguely understood. If we did not drive our cars on the right side of the road, our highways would not be safe. Without laws and authorities we would have chaos. When man is at his most antisocial, he is waging an organized war against members of his own species. No other living creatures engage in such folly. But man has learned the survival value of community, family ties, play, and communication. Outlaws and hermits are in the minority.

> "Whether for reproductive or protective ends, social behavior is essential."

We can see the value of social relationships in nature. The group behavior of dolphins and schooling fish can be considered as a social relationship because some form of communication seems to exist between these animals, and they derive benefits from their association. Territorial, aggressive behavior is social (and positive), even though it seems to be just the opposite. A fish, like the familiar grouper which marks out a territory and defends it against invading groupers, is acting to protect his species. The grouper population is dispersed over a large enough area so food supplies are ample for each individual and are not overharvested. The grouper that would usurp another's territory is warned away by the ritualized aggressive behavior of the resident. The meaning of the gesture is clear to both; the outcome of the dispute is decided without a debilitating battle or a fight to the death.

Most animals that reproduce sexually must have at least brief contact with one another, if for no other reason than for sperm to fertilize eggs. Whether it is for reproduction or for protection, social behavior in animals is of positive survival value.

As was discussed in the volume *Instinct and Intelligence,* the inventory of behavioral acts is a result of the combination and recombination of the basic drives, such as those for food, survival, or reproduction. As animals become more complex, the possibilities for intricate behavioral patterns increase.

Rather than approaching this subject as a behavioral psychologist and isolating each aspect of a behavioral act in a laboratory, we will look at some marine animals under normal conditions in the sea. The origins of social interactions will be more difficult to elucidate but at least we can assume that what we see is normal. Studying animals in an abnormal environment—the laboratory—can be compared to making generalizations about the behavior of an animal in a zoo to describe the behavior of the animal's society at large. Recently three Nobel prizes were awarded to Lorenz, Tinbergen, and Von Frisch for their studies of animals in the wild. These scientists are called ethologists; they are pioneers in a fascinating new field.

Two sergeant majors appear to kiss. *The action may be part of a mating ritual that will trigger egg laying and sperm release.*

Flashy Dresser

An individual's survival is often due to different factors from those that determine the success of an entire species. If an adult does not reproduce, its life cycle is not impaired. But if all members of a species cease to produce offspring, the species is endangered. The marine environment poses unique reproductive problems, one of the most critical of which is locating a suitable mate in the vast, dark reaches of the sea.

The underside of a midshipmanfish *(top) is lined with photophores—cells that produce light by means of chemical reactions. The glow from the photophores (bottom) assists fish in locating mates.*

One of the "brightest" methods that a marine animal has developed for sexual attraction is the flashing-light show that male and female midshipmanfish display in mating. Strung along their bodies, like sparkling midshipman's buttons, are lines of photophores—luminescent cells.

In order to examine the mating behavior of midshipmanfish, several pairs were observed in the laboratory. By using injections of adrenalin to stimulate the production of light from the photophores, the light show was artificially induced. The female gave off a generalized glow due to the reflection of her light off the sand. Once the male was attracted to the female, he flashed his photophores off and on in a blinking display lasting about two seconds per flash. Then mating took place.

Such displays serve at least two purposes. First, the two sexes are visually attracted to each other. Second, the light show serves notice that both animals are sexually receptive. The courtship display of these fish is a specialized example of sexual selection. Those animals that are the most noticeable are more likely to mate. Thus the sexually attractive trait, photophores in this case, becomes more highly developed over a period of many generations.

A Man Around the House

Childhood and youth are hazardous times in the sea. Maturity brings a stronger ability to resist disease and predators. Experience teaches even lower animals life-saving lessons. In the sea, where hiding places are few and predators many, offspring are given a better chance of survival by parents that incubate eggs in their mouths, like two species of Bahamian jawfish. Often the male plays the maternal role.

The male marine catfish retrieves fertilized eggs and carries them in his mouth. He is not content to relinquish the young once they are hatched, sheltering them for as long as two weeks after birth. The male *Apogon imberbis,* a Mediterranean fish, incubates over 20,000 eggs in his pharynx. He engulfs the eggs and constantly swirls water around them to keep them aerated.

A small crustacean, *Phronima,* eats away parts of salps to form a gelatinous house. The eggs are laid in the barrellike construction, and the animal swims about, pushing a house full of young in front of it.

__A gelatinous house__ (top) constructed from a salp becomes the brood chamber of a crustacean, Phronima. ***The home of the jawfish*** *(bottom) is an underground den fashioned from encrusting sand.*

In many animal societies the responsibility for protecting women and children falls to the male. The toadfish is an example. The female lays her eggs in a nest built under a rock, in an old tin can, or on eelgrass. During the three-week incubation period, the male guards over the nest, repelling intruders.

Woman's Lib

Sex reversal is common in many marine animals. It was thought to be a regular pattern of the life cycle of many reef fish. But recently a thought-provoking discovery has been made—in one wrasse species the sex change from female to male is due to social stress within the wrasse community.

The behavior of the cleaner wrasse, *Labroides dimidiatus,* was studied extensively on Australia's Great Barrier Reef by D. R. Robertson from the University of Queensland. The typical family of this wrasse is headed by the oldest, largest male who dominates a harem of three to six females. Each female has her own territory within that of the male's. One female, usually the largest, dominates all the others from her position in the center of the male's property. All feminine behavior is much more passive than that of the male's.

The family head makes daily trips around his territory to visit all the females in his family and to check his borders to make sure that neighboring males aren't invading. A wrasse community has been known to function in this stable fashion for as long as two years. If lesser females die, their property is annexed by other females or taken over by a young female. If the dominant male dies, the entire family undergoes an immediate shakeup. Within two to three hours the dominant female becomes less sedentary and more aggressive. She begins to initiate male aggressive displays as she makes visits around the territory. If she successfully stands off neighboring males, who rush in to capture the property of the deceased, she will become a full-fledged male!

Within one to two days, male behavior patterns are solidified. The switch to courtship and spawning behavior takes a little longer.

It is complete within four days. The "new" male soon develops the ability to produce sperm and is, indeed, a male in every respect.

The fact that the changeover is regulated by lack of male dominance is shown in the behavior of a female that fails to win the territory from invading males. She becomes a he, then reverts to a she. In one case a female that had reverted to being a female again became a male when her dominance overcame a weaker invading male.

The male ***wrasse,*** Labroides dimidiatus, *dominates a harem of three to six females. The death of the male stimulates the transformation of the senior female into a fully functional male.*

All females of this wrasse variety have the capability of turning into males. Male sexual organs are latent in each. Only one female makes the change because the intricate social structure of dominance between females ensures that she is the only remaining member of the family that does not receive aggressive treatment once the male dies.

Dr. Robertson believes that the production of males only when they are needed is of importance to the survival of this fish. The male mates with many females. His characteristics will appear in more offspring than those of any one female. He must then be genetically sound, or aberrations could be introduced into the community. By producing the male from the oldest, best-adapted female, male genetic superiority is assured.

Living Space

in some ways the housing situation in the sea is much the same as that on land. In reefs and coastal areas where food is plentiful and space is in high demand, animals pile one on top of the other, just as human beings do in large metropolitan areas. In the wide-open sea, where food is in short supply but space easier to come by, fewer animals are spread over larger areas.

Every living thing needs space. Large nomadic animals claim thousands of miles of sea as their range. Sessile organisms, those that stay put in one place, need a much smaller space but still demand a personal area. Some like crabs and starfish find their space by floating aimlessly until a certain stage in their lives when they sink to the bottom. In this fashion, their populations are dispersed over a wide area.

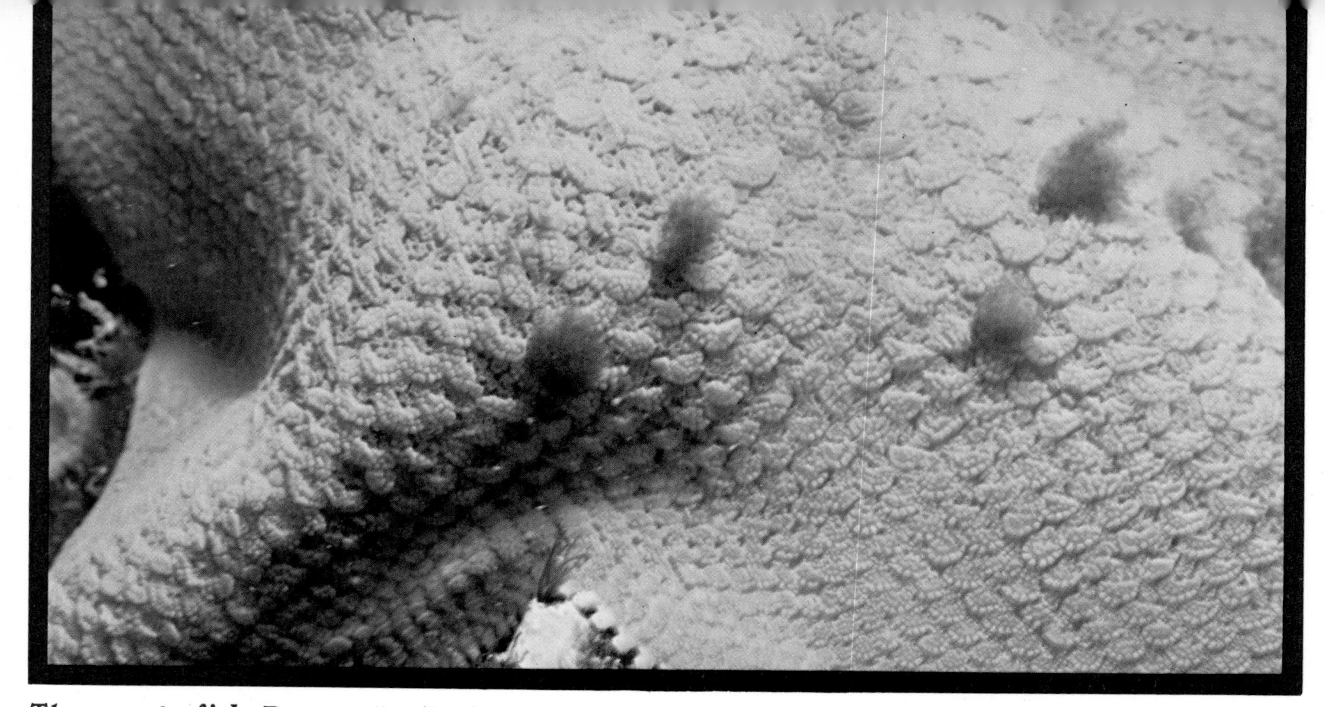

The ***sun starfish,*** Pycnopodia *(left), moves next to a red abalone. Respiration is carried out over the starfish's body surface.*

A close-up of the sun star (below) shows it is covered with white ***pincerlike appendages*** *which prevent organisms from settling on it.*

The ***bat starfish,*** Patiria *(above), has no pedicellariae and so must sacrifice some of its respiratory surface to tufts of algae.*

One species can crowd out another as well. Most starfish and urchins keep their bodies free from "settlers" (encrusting organisms) by the action of pedicellariae. These tiny, pinching organs are found over the upper body surface of the animal. When larvae, small animals, or plants settle on the back of the starfish, the pedicellariae grasp them before they can become established, and keep the starfish's body surface clean.

Any novice diver who has rested on the bottom and allowed his bare arm to touch a starfish can attest to the effectiveness of this method of defense. The hairs on the arm become engaged by the pedicellariae and are pulled out when the starfish is removed.

Starfish aren't vain, wishing to keep themselves free of algal growths for aesthethic reasons. Pedicellariae have evolved for respiratory reasons. Starfish "lungs" are protrusions of the internal lining through the surface skeleton of the starfish—they breathe through their skin. Tiny cilia keep water circulating around the "lungs." If the animal becomes overgrown with fouling substances, the breathing mechanism is impaired. Surprisingly some starfish do not have these pincers and support tufts of algae, while some urchins with pedicellariae have barnacles growing on their backs.

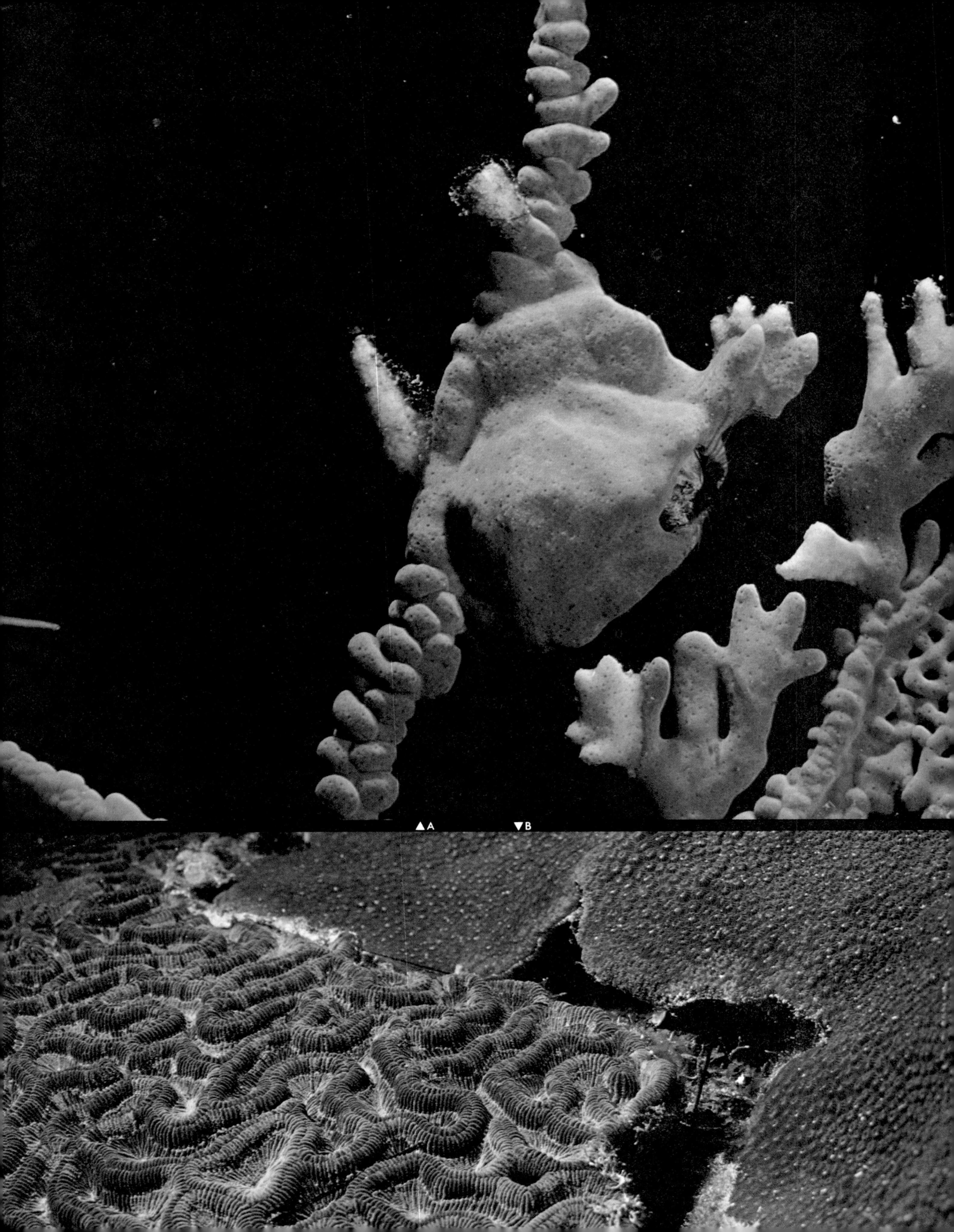

▲A
▼B

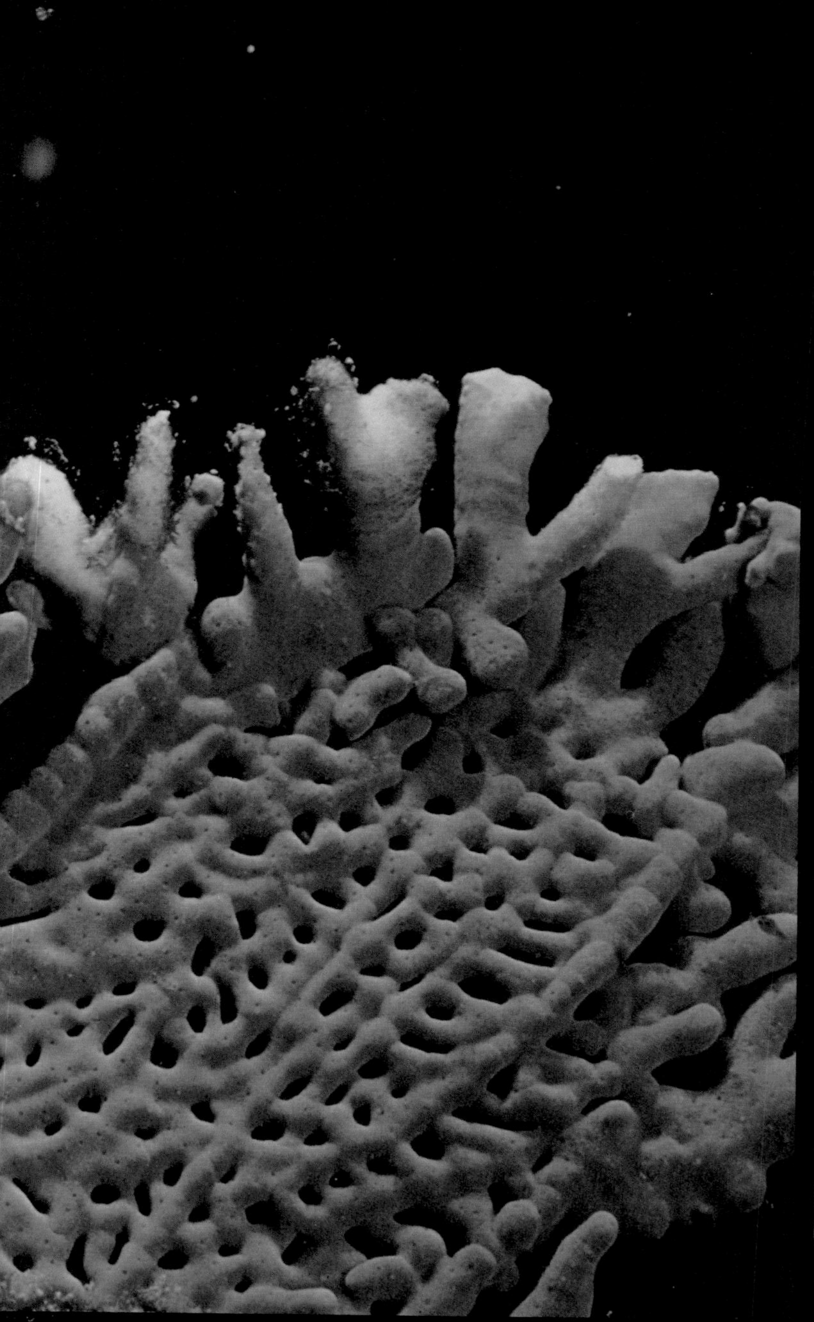

▲C ▼D

Competitive Corals

Living space is at a premium on reefs. Many corals are antisocial, and compete for places to live. Some grow over and encrust members of their group, while others coexist, exhibiting zones of interaction.

***Fire coral** (A) has completely taken over a sea fan, using it as a living substrate. As corals grow they meet and can identify members of another species (B, D), or even members of the same species (C), but with a different genetic heritage.*

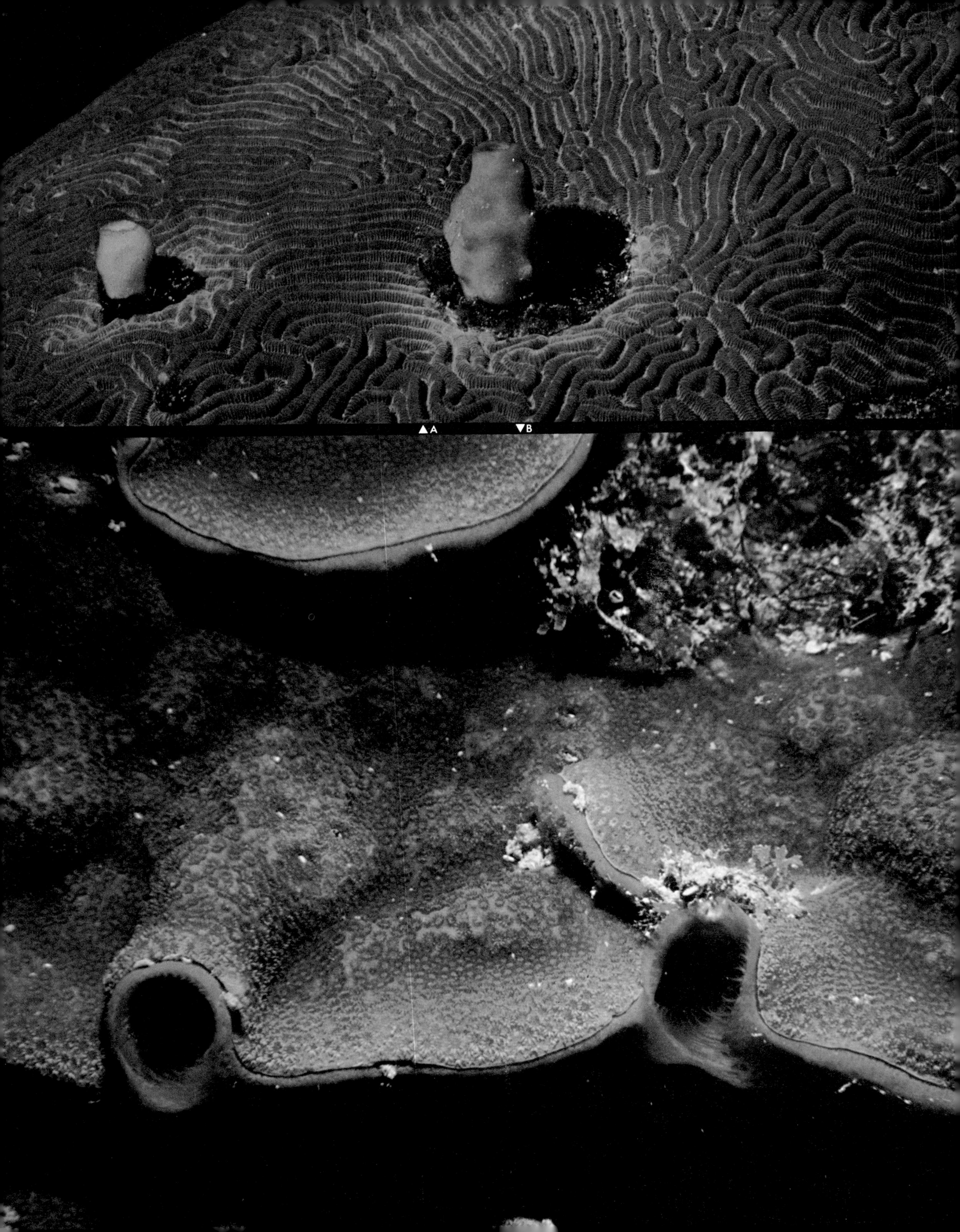
▲A
▼B

▲C

Silent War

Even though corals have effective defenses against some animals, they are subject to invasion by others. Certain species of sponges may be considered antisocial and can bore into corals, possibly utilizing them as a foundation for living space.

Two ***sponges*** *(A) have excavated holes in a large brain coral, thus providing for themselves a rock-like foundation to live on. An orange sponge (B) extends out from below a coral, possibly inhibiting its growth. Other yellow and orange sponges (C, D, E, F) invade and destroy reef-building corals.*

▼F

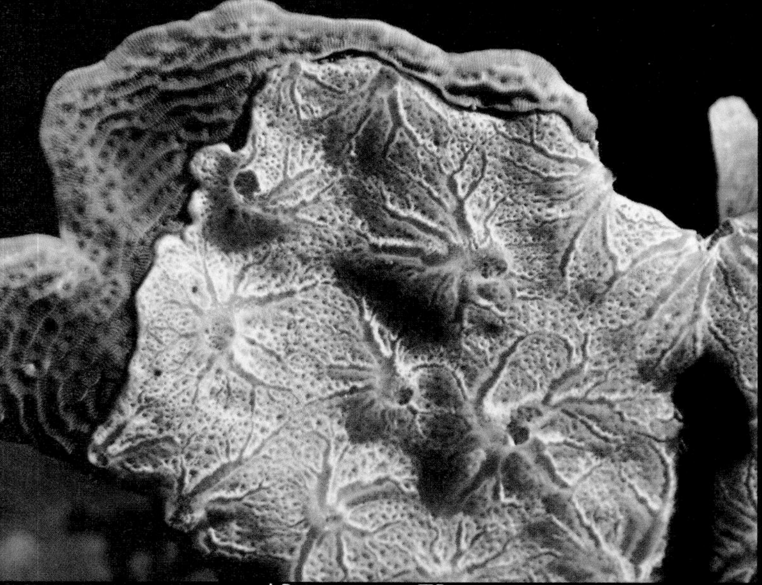

▲D ▼E

Living Together

In some cases sponges are antisocial, living on coelenterates (corals) and boring into them, causing considerable damage; in other instances the relationship is more social, with the coelenterates (anemones) residing on the sponge. Although the filtering surface of the sponge is reduced, superficial observation reveals that the coelenterate has no damaging effects upon the sponge.

Shown on these pages are ***Caribbean sponges*** *infested with epizoic zoanthid anemones.*

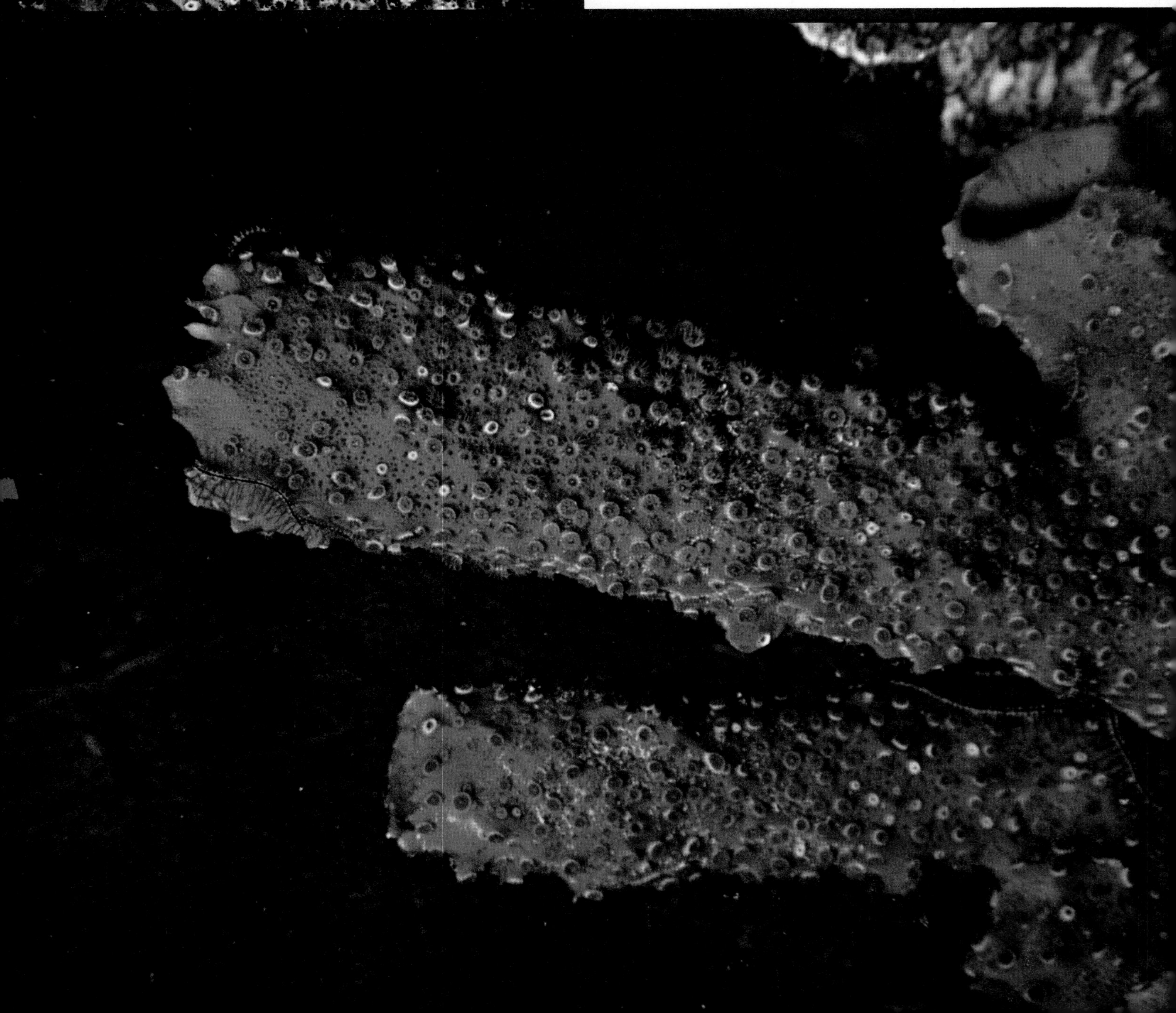

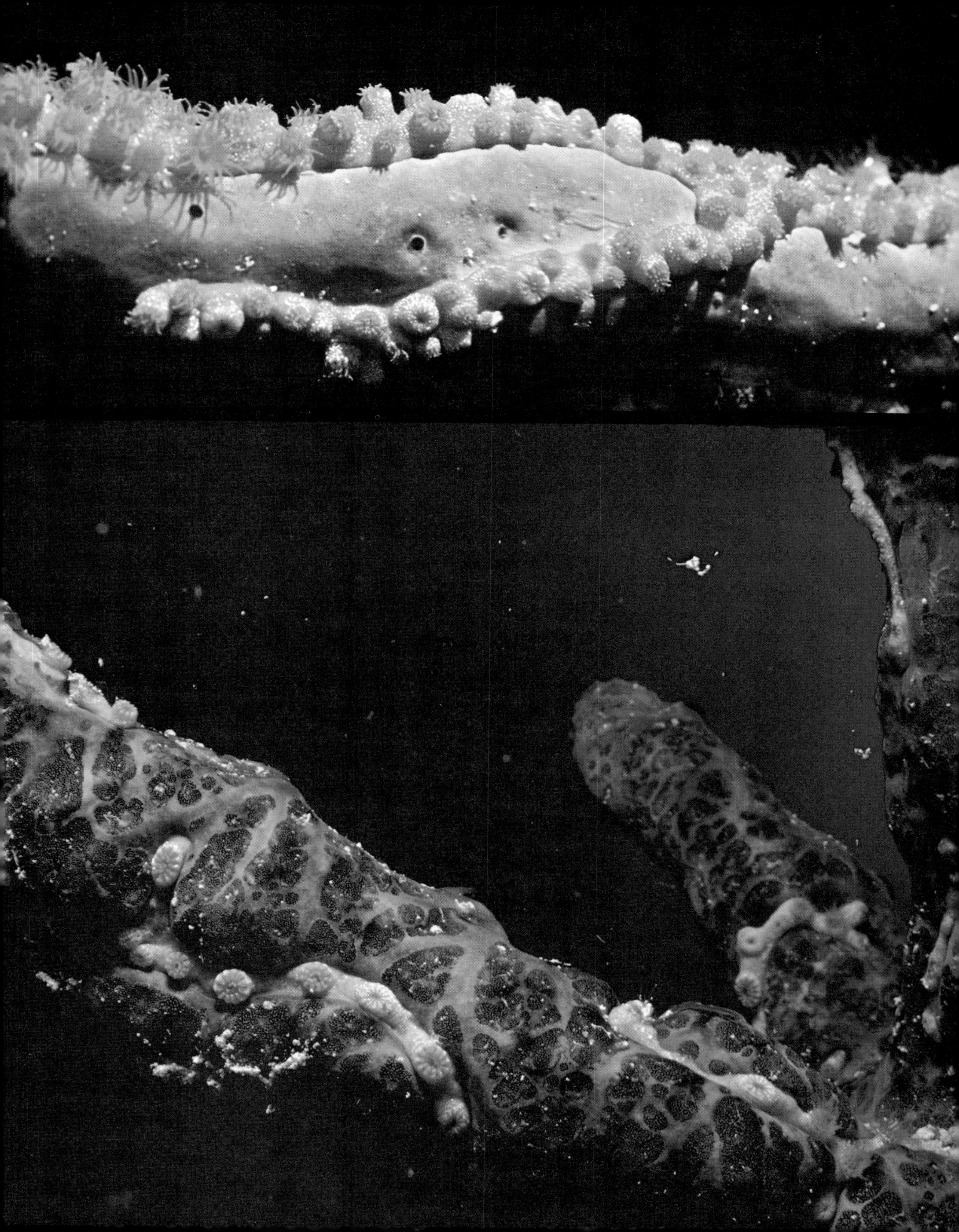

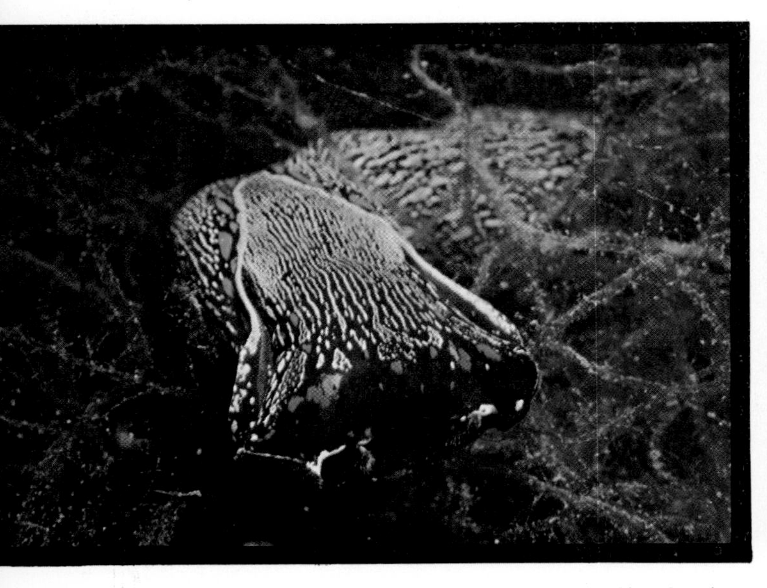

Navanax *(above) senses the mucus trails of other invertebrates with its small tuftlike receptors.*

A small Navanax *has been put on a* ***collision course*** *with a larger one (above). The smaller one lunges*

Cannibal or Lover

Social behavior in lower animals can be highly intricate, but it is generally much easier to decipher than that of more complex organisms. The day-to-day life of a primitive animal revolves around feeding, survival, and mating: its nervous system is not usually complex enough to allow for a broad spectrum of social interactions.

Navanax, a type of sea slug that can be considered a "simple" organism, demonstrates both social and antisocial behavior toward its fellows. Unlike mammals and other higher animals, the behavior of *Navanax* is based solely on one-to-one interactions triggered by the senses of smell and taste.

Navanax is commonly seen in muddy bays and inlets along the coast of southern California. Its life revolves around a constant search for other similar molluscs upon which it feeds. Although *Navanax* has eyes, it relies almost completely on chemoreceptors to locate its prey. Two prominent, elbowlike structures that protrude from the head of *Navanax* are used in locating mucus trails left by other snails and slugs. Hairs on the protrusions contain organs that are sensitive to the taste and smell of mucus trails. *Navanax* is so dependent on its sense of taste to locate prey that it will unhesitantly swallow pieces of paper or artificial sponge which have been soaked in the mucus of a prey species. It is also attracted to the smell of the mucus trails left by other *Navanax.*

One *Navanax* will unerringly follow the mucus trail of another. When one overtakes another, the larger *Navanax* may swallow the other one whole in a quick gulping action. In an encounter between two *Navanax* of equal size, perfunctory bites may be exchanged. Once it has been determined that predation is impossible because of their matched size, the two animals may mate and then go their separate ways.

Some shell-less relatives are protected from predation by the *Navanax* because they are large in comparison or because they produce acidic mucus trails. Although *Navanax* will follow the mucus trails of an acid-producing species and swallow the animal, *Navanax* immediately spits it out.

at the other (above), possibly forestalling an attempt by the latter. Then they separate (above right).

Sometimes, a larger Navanax *will* ***attack and eat*** *a smaller one when contact is made (below).*

Another method of protection against *Navanax* exists among molluscs that inhabit the same areas. One species of nudibranch which has nematocyst-charged cerata on its back reacts dramatically to *Navanax* mucus. Upon contact with it, the cerata elongate and thrash about violently, releasing the nematocysts. This reaction sometimes, but not always, causes the *Navanax* to withdraw and set off on an alternate course.

Self-Control

Some species appear to practice self-limiting behavior in order to keep their populations under control. Some branches of staghorn coral suffer from "white death," in which branches die from the tip to the base. Since no parasite or disease seems to be the cause of this action, it is thought that the coral is practicing self-pruning.

Within a short time after a branch has died, algae and fungi colonize the area. Soon parrotfish eat away the infested area. After a few months the branch is completely gone; a healed scar is all that remains.

Since crops of staghorn coral produce far more branches than are needed for survival, unneeded ones are sacrificed. If this did not occur, an overgrown tangle would result. Water circulation and light penetration would be inhibited to all but the coral branches on the outside of the tangle.

The phenomenon of white death can be observed in ***staghorn coral.*** *It is a self-pruning technique that prevents the colony from choking itself.*

Losing Yourself

Living together with many different species is a constant fight for survival between weak and strong. But the most vicious of carnivores is not always the best equipped to wage the battle for life in the sea. The timid creatures too have defense mechanisms that are less fearsome but equally effective. The process called autonomy is a subtle but effective method of escape.

"Autonomy" means "self-cutting." Animals that possess this ability do just that, letting go parts of their body when threatened.

Autonomy is an interesting anatomical feature found in some starfish, a particular type of file shell, and the jackknife clam as well as others. A sea snail with a reduced shell, *Oxynoe panamensis,* lives in shallow mangrove swamps in Baja California, where it feeds on algae. When the tail of this gastropod is pulled on, it comes off twitching while the animal makes its escape.

A brittle starfish, *Amphiodia,* has long, skinny arms that it extends up from the bottom to gather food. If it is dug from its buried home, three or four inches beneath the mud, it throws off parts of its arms, one by one.

The tentacles of these ***file shells*** *(open and closed) are autonomous. If a predator should grab them, they come off, leaving the file shell unharmed.*

Chapter III. Facing the Elements

Storms sweep across the sea, leaving destruction in their path. Fragile animals lose their moorings as huge waves sweep them off their rocky homes. Salt water rushes up freshwater estuaries, killing salinity-sensitive organisms. Or the reverse occurs: heavy rains pour down on land, swelling inland rivers and bringing fresh water far out to sea. Hurricanes stir up sandy bottoms, burying helpless animals in silt that clogs their bodies.

"No marine habitats are more vulnerable to the challenges and forces of the elements than the tidal and coastal areas."

Occasionally even more violent eruptions occur. Earthquakes alter the face of the coastline itself. Volcanoes bubble and spew up from the bottom, creating new islands and destroying existing environments.

No marine habitat is more vulnerable to the challenges of the elements than tidal areas. Waves continually pound at the shore, during good weather and bad. Water, the lifeblood of marine creatures, slowly moves out of their reach as the tides ebb. Temperatures vary greatly as the sun beats down on a life-encrusted rock that was bathed in cooling water only hours before.

The weather on land is not the same as for adjacent bodies of water. The temperature of the sea changes slowly. A cold front that passes briefly across a coastline is much more noticeable to land life than to that in the sea. The organisms that live in the tidal zone must endure both extremes. When the tide is out, they are exposed to terrestrial weather. Once the sea enfolds them again, they are in the marine environment.

In the face of such daily periodic fluctuations, the only creatures that can endure this environment are those with prodigious reproductive capabilities. Animals that can reproduce quickly can repopulate areas where populations have been wiped out by the elements. Some snails and limpets attach so firmly to the rocks that virtually no water is lost at low tide. They are aided by a watertight mucous secretion. For the animals of this border, daily environmental changes are just episodes in the never-ending struggle between life and death.

The beach is the slate upon which the story of a storm at sea is written. The morning after a violent surf or heavy rain is a beachcomber's delight. The remains of animals that live beyond the range of the tides are brought to the land. Seaweed and shells of deep-water clams are found scattered among the flotsam. Floating animals are often driven to land by storm winds. Every few years great numbers of "by-the-wind-sailors" jellyfish are washed up on the California coast. Most are but shadows of the beautiful creatures that they once were before catching a ride on the storm, but occasionally a perfect specimen can be picked up on the beach.

In the North Pacific communities of the United States, a popular hobby is based on collecting glass floats from Japanese fishing nets. Storms bring these objects in from the sea, and hundreds of children go to the beach in their pursuit. Even today, when the water barrier between countries has been shrunk by air travel, it is hard to resist poking among the debris on a beach for a remnant of another culture.

The surf is a hostile environment. *Waves pounding along a shoreline prevent all but the hardiest plants and animals from inhabiting the area.*

A rocky shore, *exposed at low tide, shows a clear zonation. Species that are best equipped to withstand desiccation are found in the higher zones.*

Split-Level Housing

Every animal has a niche. An ecological niche is defined as the physical location occupied by an animal, together with the food that it consumes and the other animals with which it interacts. The environmental factors that influence the kind of organism that can succesfully populate an area vary greatly and give both a horizontal and vertical zonation to the shore. The most graphic illustration of ecological niche is found in the stratification of animals and plants on an exposed rocky shore. Each species is found in its own horizontal hideaway.

Closest to the high-tide mark cluster one or more species of barnacles, segregated horizontally by species. On the rocky coast of Scotland the barnacle *Chthamalus stellatus* occupies the highest zone. Beneath it are usually found colonies of the barnacle *Balanus*. Since only young *Chthamalus* are found beneath or in with *Balanus,* scientists attempted to explain barnacle zonation in terms of crowding by the fast-growing *Balanus*. All *Balanus* were removed from an area where young *Chthamalus* had settled. In the absence of the faster-growing barnacle, the young of the other species were able to establish a strata beneath their normal one. The *Balanus* crowds out and crushes any other species that attempts to colonize in its area.

But why don't the faster-developing barnacles also push their relatives, *Chthamalus,* out of the upper tidal regions? The answer is that *Chthamalus* is a hardier species that can resist the longer periods of dryness and exposure near the high-tide mark. *Balanus* cannot survive these harsh conditions.

Beneath the barnacles are usually found mussels which occupy this stratum because they cannot tolerate prolonged exposure to air. Their strong byssal threads enable them to cling to exposed rocks receiving even the greatest surge. The lower limits of the mussel are defined by the upper limits of their most dangerous predator, starfish, which effectively remove them from those low tidal levels in which the starfish can survive. The starfish are even more sensitive to exposure than the shellfish and cannot venture too high above mean tide in search of mussels.

Dense mats of vegetation are partially exposed at low tide. ***Tide pool flora*** *must be able to survive wide variations in temperature and salinity.*

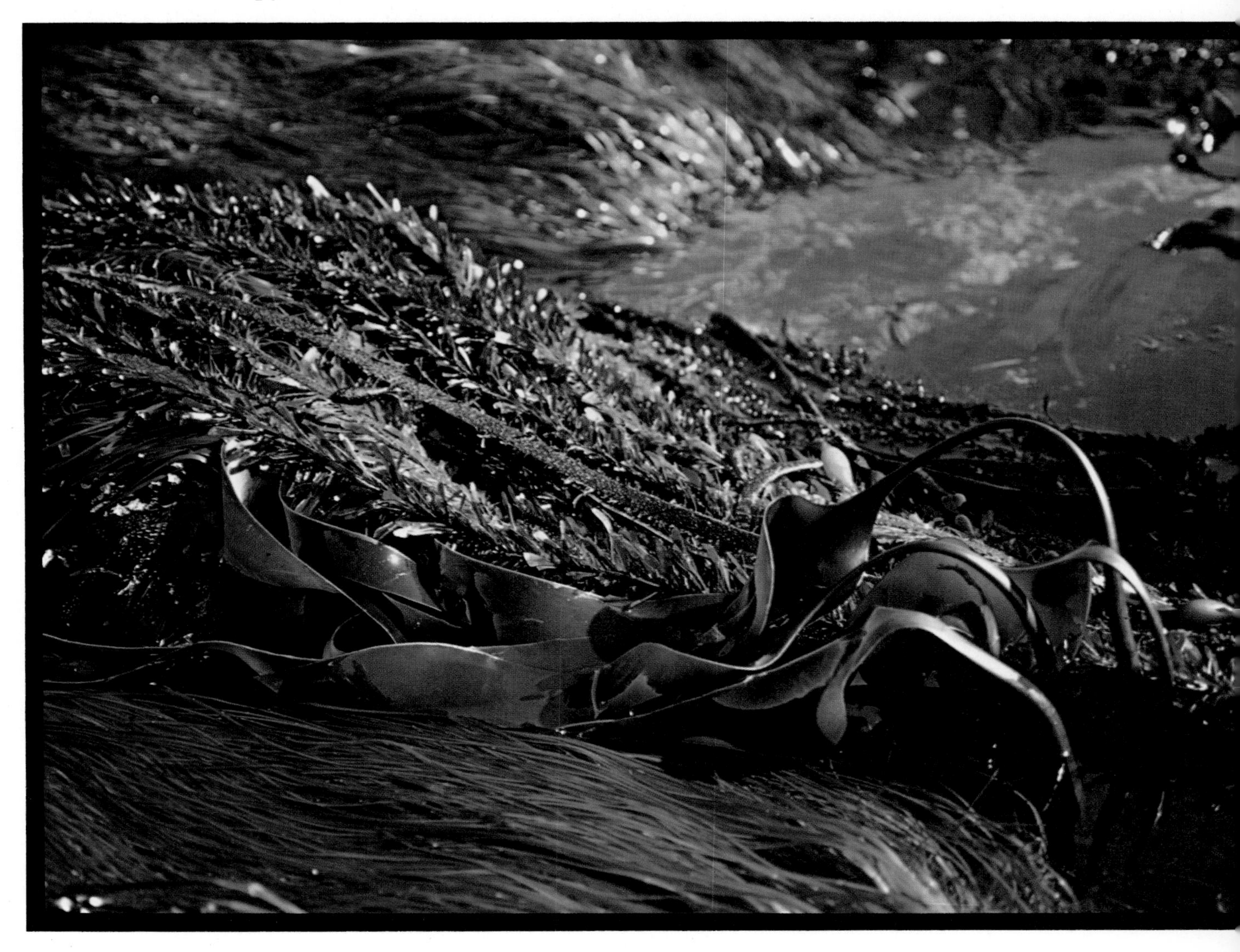

***Shelled molluscs** (chitons) cling tenaciously to a surface (below). A partial vacuum under their feet makes them almost impossible to dislodge.*

*A **mass of gooseneck barnacles** (above) firmly attaches to the substrate. The mouth and feeding appendages are at the extremity of the stalked body.*

A Permanent Base

Life on a rocky coast requires both physical and behavioral adaptations in order to withstand the force of the battering waves—up to two tons per square meter. Most rocky shore organisms solve this problem in a similar fashion: once they land on a suitable site, they cement themselves to a surface, so that the waves cannot carry them out to sea or to death against the rocks.

Barnacles attach themselves permanently to the first rock they land on. The ability they have to survive the battering waves until they get their mooring is amazing. Once attached, they can't forage for food and depend upon the waves to bring them sustenance. The way a barnacle feeds has been likened to an animal that lies on its back and kicks its food into its mouth with its feet. Barnacles are crustaceans, and the filaments they use to filter plankton from the water are derived from the structures that develop into legs in other crustaceans. When the tide ebbs and the barnacle is high and dry, another adaptation becomes of critical

*An **intertidal community** is complex. Old and newly settled acorn barnacles, common limpets, and rough periwinkles have invaded a rock surface (above), while common mussels fill the crevices.*

importance. In acorn and gooseneck barnacles, the aperture through which the feeding structures extend is covered by movable plates that close over the opening to protect the animal from drying out.

Barnacles are hermaphroditic—both sexes are contained in one animal. Even so, two organisms are needed for sexual reproduction to take place. Since barnacles can't move from their position to mate, they must be anchored close to other barnacles so that the sperm-carrying tubes that extend from one barnacle can fertilize the eggs contained in another and vice versa.

Barnacle larvae appear to be attracted to established barnacle beds by chemical messages. They do not often come to sexual maturity on an isolated rock far from a mate.

Beachcombers

To the casual observer walking on a beach at low tide, the gently sloping sand seems void of life, especially when compared to the lively tide pools on rocky beaches. Life is present, but it is in and under the sand.

The primary producers (plants) in the beach ecosystem are also not evident to the walker. Occasionally a bladder kelp washes up to give indication that large plants grow offshore, but for the most part the plants of the sandy shore are microscopic plankton that wash in with the tide. When the tide is out, the basic food source of the beach is removed.

Comparison of the number of species that inhabit the sand with those that live on the rocky coast shows that the two habitats support an equally diverse population. The lack of food on the sandy shore limits the population of all species, however. Almost all organisms that find sand an appropriate home are burrowers.

Some creatures, like the sand crab, alternately burrow to prevent being washed out to sea and then jump up to forage about at low tide. The tendency for sand to cave in makes permanent burrows on the sandy beach a difficult design problem. *Pectinaria,* a marine worm, constructs its tube by cementing a single layer of fine sand grains together with mucus. The workmanship of the tube is precise; each grain is lined up in perfect order. The construction is so sound that empty tubes sometimes wash up on shore intact.

Since the incoming tide brings food to the shoreline, those animals that can follow the leading edge of the sea are obviously best equipped for shore life. A crustacean of the amphipod family, *Corophium,* burrows into the sand and draws food-carrying water through a horseshoe-shaped tunnel.

The beach hopper *(below) is an amphipod* (Crustacea), *often found swimming in beds of algae.*

Wave action has torn fronds *of a brown algae (above) from an underwater garden.*

Trails left by snails *(below) are the only visible traces of animal life on a beach at low tide.*

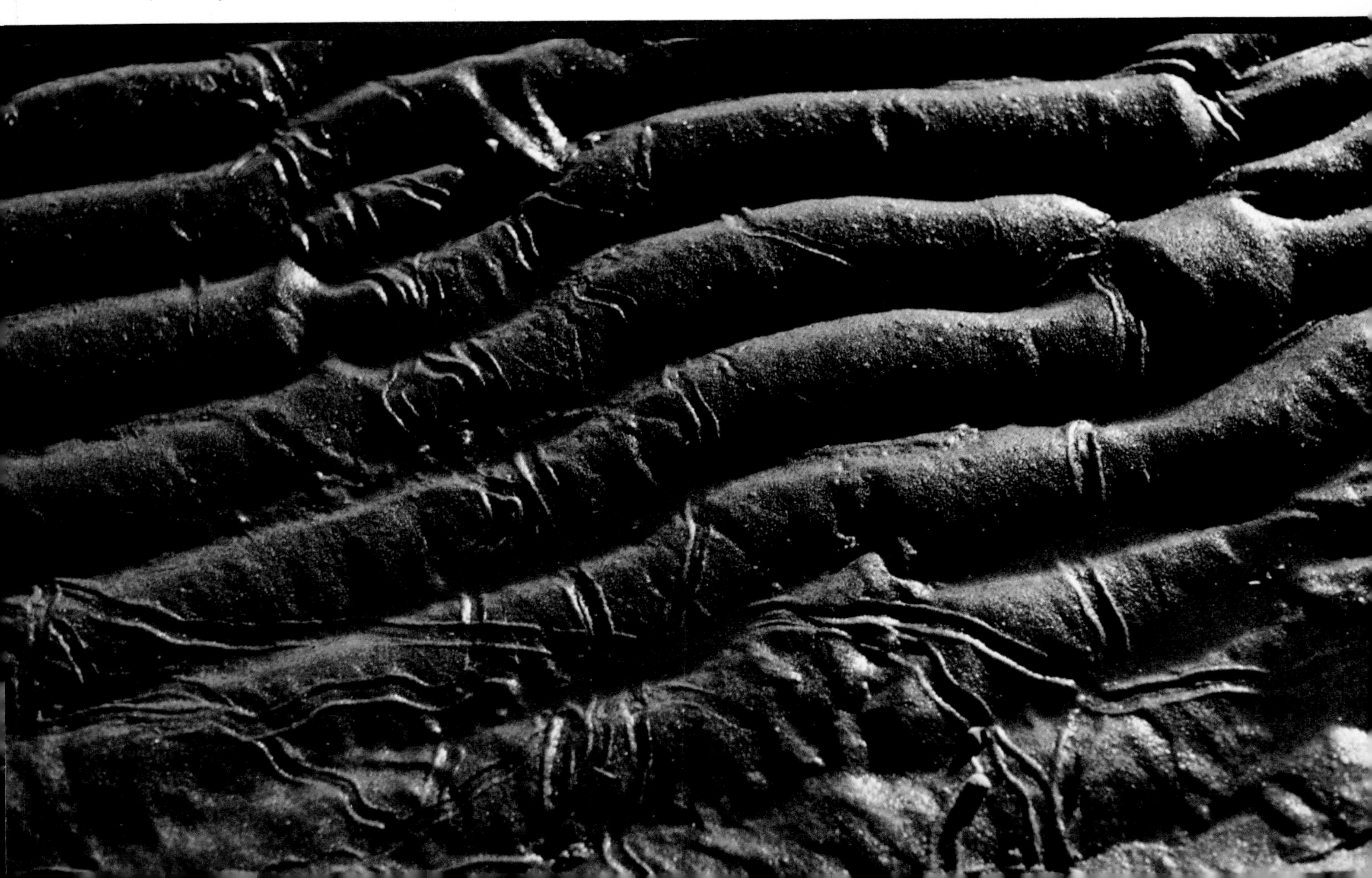

A seaward view across algae-covered mud flats illustrates one of the three ***feeding habitats*** *frequented by* Hydrobia. *If fresh water flows out over the flats,* Hydrobia *will burrow under the mud.*

Living in Two Worlds

A freshwater amoeba placed in salt water will quickly shrivel up and die as its water content responds to osmosis—the movement of molecules through a membrane from an area of higher to lower concentration. Similarily, an animal that lives in salt water becomes waterlogged and dies as water diffuses into its body from a freshwater environment. Many animals have internal "pumps" that enable them to defy the laws of osmosis and tolerate a salinity range from fresh to salt. These animals and plants are most successful in estuary habitats where fresh water moves into the sea from rivers and the tides carry seawater up into streams.

Hydrobia, a tidal snail that has been studied extensively in British estuaries and beaches, is seemingly perfectly adapted for the rigorous life in estuaries, both from a physiological and behavioral point of view.

We can begin the story of *Hydrobia*'s day with low tide. If it is not new to the beach or a bank, it is probably buried in the sand to avoid desiccation. However, the snail may be roving about the sand or mud to eat, deriving nourishment from bacterial films that cling to sand particles. As the tide comes in or the river swells, *Hydrobia* constructs a raft out of mucus upon which it sets sail upstream with the tide. Plankton and organic material are entrapped in the mucous net, thereby completing the snail's feeding habits.

The snail is successful since it can feed at both low and high tidal levels.

The danger involved in moving upstream is that as the snail gets farther away from the sea, salinity begins to decrease abruptly. Studies on *Hydrobia* in an estuary where high tide was some eight feet above the river-bank revealed that the snail can tolerate salinity variances from 2.6 parts per thousand to 15.4 parts per thousand.

When the water is almost fresh, *Hydrobia* halts its passage farther upstream by closing its operculum and sinking. Thus low salinity stimulates a behavior response that protects the animal from being carried too far from the sea. If the salinity falls too low, as when fresh water goes far into the sea, the snail does not initiate the floating aspect of its feeding habits and remains in the lower reaches of the estuary where the salinity is still within the snail's range of tolerance.

This animal's tolerance for a broad range of salinity concentration appears to be based on two factors. First, the animal apparently has a "salt pump," which allows it to regulate the salt in its blood, and thus it is capable of living in fresher water. The pump works so rapidly that the animal does not have an opportunity to take up excessive water. Second, the snail can withdraw from the water by sinking. If conditions become too intolerable, the animal burrows into the sand to await better times. Diversity of behavior responses makes *Hydrobia* highly successful.

Numerous Hydrobia *are* ***stranded above the mud flats*** *during the low tide. They close the operculum to prevent desiccation and await the tide that will carry them upstream to feed.*

Dollars in the Sand

The degree of wave action and direction of water movement along a beach vary greatly during storms. Some animals succumb to the upheaval of their environment during surf. Those that have behavioral adaptations that allow them to escape adverse weather conditions have obvious advantages over those that don't.

An interesting relationship exists between the water movements along a coastline and the behavior of sand dollars. These creatures adapt well to the sandy, exposed coastlines on which they live. In some places sand dollar colonies form vast beds containing millions of individuals. The beds are oriented parallel to the coastline for miles.

The spines of the sand dollar are reduced to a fuzzy covering which acts as tiny feet, allowing the sand dollar to move very well on its unstable substrate. In addition to physical adaptations, the sand dollar has behavioral mechanisms that afford some degree of protection from the elements. At the innermost regions of the beds, just outside the surge zone, young sand dollars lie horizontally submerged beneath the sand where the waves cannot tumble them around. Here they feed on detritus in the sand.

A diver moving seaward from the surge area will notice that small patches of sand dollars begin to appear. Surprisingly, they do not lie flat but stand on end in the sand. At the outermost regions of the bed, individual sand dollars are so closely packed that the sand seems black. These are the largest adults. In some areas the population is limited to a precisely defined line, beyond which no individuals are found. It is not known why sand dollars are reluctant to move out beyond the bed. Some scientists studying them believe physical factors cause this behavior.

Part of the success sand dollars enjoy in their ecosystem depends upon their vertical orientation. When the water is relatively calm, all the individuals lie vertically, or almost so, with their flattened bodies at an angle to the bottom. Their oral side is down. At this stage they feed on organic particles and plankton in the water. When the action of the surge increases during storms, they tend to lie at an angle closer to the sand. Finally in a heavy storm surf, they lie flat, at times burying themselves in the sand.

As would be expected, the sand dollars are always oriented parallel to the surge so as to reduce its effect on them. In calm bays, there is no common orientation pattern—individuals lie every which way. On exposed coastlines which consistently receive heavy surf, all the individuals lie inclined, partially or wholly buried in the sand.

After storms the sand dollars that wash up on the sand are often those that are heavily fouled with barnacles or other clinging organisms. It is thought that only sick animals are colonized and that this encrustation prevents the sand dollar from practicing the patterns that allow it to survive a storm.

During storms, ***sand dollars*** *(opposite) lie at an angle close to the sand. In a severe storm they bury themselves to avoid being swept away.*

During a light surf, the majority of a colony of sand dollars (opposite, insert) ***stand on edge,*** *orienting themselves parallel to the surge of water.*

Sea pens *(top) stand at attention on the muddy bottom. Like sand dollars, these colonial animals can work their way into the sediment for protection.*

Another species of sea pen *(right) displays similar behavior. When the bottom is stirred by a storm, the sea pen retracts into an underground tube.*

Chapter IV. Life is Change

It is a paradox in nature that changes are indispensable to stability. Checks and balances between predator and prey, food supply and populations, invasions and rejections, as well as constant environmental changes, all bring stability to a community. The result is a biological community in which each animal has its role or niche.

Sometimes it seems that two very similar animals are occupying the same niche—theoretically an impossibility. Two species of cormorant, the common cormorant and the shag, share the same stretch of beach in England. They both dive into the sea for food and rest on cliffs overlooking the ocean. Closer examination shows that they occupy different niches. The shags eat sand eels and sprat (a fish), while the common cormorant eats shrimp. Thus the two animals do not interfere with each other's food supply.

"Man's solution to stagnant society is revolution. Nature's is also a catastrophic event."

If a community is destroyed, all the niches are temporarily eradicated. A slow process of replacement begins to restore the community. Pioneer species settle in the area. They set the scene for new species until a climax community is restored.

The climax situation is not at all the healthiest form of a community. Biological communities develop along the same paths as do human societies; in the beginning there is vitality and lively competition. As time passes, the most successful force out their competitors. In biological communities, holes appear in the food web as the number of species in the area diminishes. The community becomes stagnant and susceptible to invasion by predators. Stability means death. Man's way out of a stagnant society is revolution. Nature's way is to wipe out a community by a catastrophic event so it can begin anew.

An example of the necessity of change in an ecosystem is seen in the relationship of forest fires to forest ecology. The succession of life-forms (serial stages) that develops in a burned area has greater productivity and a greater energy flow than that of the climax. This increases the possibility of a high biomass—large population of animals. Burned areas of the forest are pockets of vitality. Since many species common to serial communities do not exist in the climax area, these areas serve as a reservoir from which new serial communities can develop after a climax forest burns.

For example, the golden eagle is absent from the interior of Alaska except the tundra and burned forests. The bird cannot survive in heavy timber and some think it would vanish from the forest if serial communities were not periodically started by natural fires.

A similar situation probably exists on coral reefs. Although no thorough studies have been undertaken to analyze the stability of various coral communities, climax and serial associations do exist. The reefs are affected periodically by devastating storms. After a storm new species invade the area and diversity rises. With time the ecosystem again matures to a climax community with less diversity in species. Some ecologists think the reefs will remain healthy only if they are periodically subjected to catastrophic changes.

A tropical reef *depends upon diversification of life for survival. Each creature fills a specific niche that contributes to community stability.*

The More the Merrier

The science of dealing with invasions of biological pests on forest communities has few devotees in countries where the forest is tropical; such a diverse ecosystem has the ability to resist pest invasions. A community like the tropical rain forest or the coral reef supports a wide number of species and is sure to have at least one predator capable of keeping an invading species in line.

Writer-ecologist Marston Bates likens the health of a diverse ecosystem to that of a nation with a diverse economy. A country with a single crop or sole business to support its people is always in danger of economic catastrophe. If that crop or business fails, the economy is doomed. A nation or an ecosystem with many food supplies can switch to an alternative if one source fails. The more complex an ecosystem is, the more likely it is to survive stress.

Why does one geographical area, like a reef, support a greater diversity of life than arctic waters? The theory is that more niches are available. The reef provides many habitats:

dark caverns and coral grottoes, surfaces for algae to attach to and bacteria to live on. In the arctic the climate is harsh with a limited diversity of life. The relation of ecological complexity to stability has several lessons for twentieth-century man. The tropical reef community is the most diverse of the marine habitats; the polar regions the least. When our business takes us to the arctic, we must be highly sensitive to the fragility of the ecosystem. Elimination of even one species by overfishing or as a result of pollution could easily upset the checks and balances. When fur seals were extensively killed by man, the polar bear suffered population decreases. The only alternative food source for the bear was the walrus. In a complex system, many alternatives would have been available.

Many individuals and few species characterize a northern ecosystem. ***Juvenile Alaskan king crabs*** *in a pod formation rest on the bottom (left).*

A tropical reef (below) shows ***a wide diversity of life.*** *The crown-of-thorns starfish is one of the many predators that periodically invade the reef.*

Colonizing a Lava Flow

The ideal way to study succession in an ecosystem is to begin observations on an area that is void of life. As animals and plants move in to colonize the habitat, pioneer species will be identified. If the observation is kept up for a period of years, the evolution of a climax community can be followed.

Natural laboratories in which to study succession are usually the result of a catastrophe. Burned forest land is a prime example. Areas where lava flows from a volcano into the sea are another. Both provide the ecologist unique, sterile environments to observe.

On January 13, 1960, a volcano erupted on the island of Hawaii. Lava from the Kapoho eruption was deposited over areas of adjacent land and overflowed into the sea. Within two days a new shore had built up that extended several hundred yards into the sea from the original beach. The lava flow spanned a three-mile stretch of coastline, destroying its marine communities.

The new beach and coastal bottom was ideal for the study of marine succession for several reasons. As it was not in the heavy surf that typifies most coasts of the island, underwater observation was facilitated. And the 1960 flow was adjacent to lava flows that had poured into the sea at other times in the past century. Comparisons could be made and conclusions drawn about the nature of pioneer organisms and the subsequent colonization of the flows.

Undisturbed coastal and offshore areas in this region of Hawaii are luxurious in coral growth. Normally these areas are inhabited by a wide variety of associated fauna, including many colorful reef fish. When the new coastal area was first observed by ecologists a year after the Kapoho eruption, no coral growth could be detected on the lava flow. The pioneer organisms on the emerging beaches were small polychaete worms. In areas of the coast where the lava flow had created outcroppings, the first life consisted of a matting of brown, green, and red algae mixed with encrusting coralline algae which were already firmly established within a year and a half of the eruption. Associated with the algae on the outcroppings were several varieties of marine snails, tube worms, and one species of crab.

The offshore bottom was strewn with loose stones and silt. Within a year sea urchins, brittle stars, and gastropods had all returned to the bottom. However, life was absent from the water above the silt even by May 1961 (14 months after the eruption). It was not until later that summer that some reef fish returned to the waters.

A new tide pool was also created by the geological havoc. In this area which the sea reaches only at high tide, a dozen species of fish were observed to be colonized in the spring of 1961. Shrimp were the dominant crustaceans in the tide pool. Several snails and their relations were also present.

The biologists who began the study of the Kapoho site in 1961 have continued observation of the area under the direction of Dr. Townsley of the University of Hawaii. After 13 years of study they concluded that complete repopulation of all species including coral will be accomplished on the flow by 1975. Judging by older, adjacent lava flows, community balances and climax structure will not be stabilized for 35 to 80 years.

***Cinder cones and volcanic ash** (left) support sparse vegetation on the Galápagos Islands. Algae, invertebrates, and **reef fish begin to recolonize a Hawaiian area** inundated by lava (below).*

New Towns

Succession and development of a marine community are dependent on many variables. Obviously, if the temperature or salinity or any other environmental factor is not appropriate for an organism, it will not colonize an area. Often more subtle variables determine to what degree populations will inhabit an area and the patterns of distribution of organisms that will emerge.

It is generally accepted by ecologists that distribution of animals in an ecosystem is a result of one of five processes. The first is *environmental,* as we have just mentioned. Or it may be *reproductive:* an animal may stay in an area to remain close to its parent. Other species are drawn into new localities by *social interactions,* such as signals that call them to the location of mates. A predator comes into a new habitat *seeking food.* Sometimes life appears for none of the above reasons. Ecologists call this a *stochastic distribution*—the result of random factors.

To define which processes are at work in development of a marine community and what influences the distribution of animals, Edward W. Fager at the Scripps Institution of Oceanography placed one-meter cubes of asbestos and iron in the sea at a depth of 12-14 meters near La Jolla, California. The cubes were designed to provide a substrate for sessile organisms and had compartments in which animals could hide. The first cube was placed in position in January 1968, the second in October, and two more in December.

Diving scientists observed the cubes about once every three weeks and noted what organisms were living on or in them. Six months after the first cube was positioned, brown algae was observed on the tops of the cubes. The pioneer animals were the barnacles *Balanus pacificus* and *B. trigonus,* which appeared in less than a month. After six months the algae population diminished as a result of grazers. The compartments in the cubes were usually occupied by large crabs, spiny lobsters, and an occasional octopus.

The boxes were in the path of currents that brought larval forms to colonize or to provide a constantly replenished source of plankton for food. The food web that emerged from the study indicated that the barnacles were, at least initially, the primary path by which energy was carried to the major predators and made available to the community.

As time passed, more and more species became associated with the boxes. Over a hundred organisms were eventually found on or in the boxes within two years of the study. Twenty-two species of fish were seen around the cubes. Apparently they were drawn to the areas to feed. Smaller fish were probably first brought to the cubes in large balls of detritus kelp that were continually swept into the boxes by the currents.

Dr. Fager's study is significant because it illustrates the complexities of a community concentrated in a few meters of space.

Tropical marine life completely hides the structure of a ***sunken ship*** *(right) which serves as an artificial reef. Cement blocks provide a good substrate for increasing life on a sandy bottom (left).*

Give-and-Take

If a new species is introduced into a stable ecosystem, native species will suffer if the intruder competes more efficiently for the same food or space. And if the new species does not appeal to predators, it will soon take over the niche of a native species and force it out.

In 1945 a barnacle that is native to Australia appeared in a few spots on the southeast coast of England. By 1947 it had spread out from these areas and is presently found all along the west coast of Europe and even down the coast of Africa. It is thought that the barnacle came to England by traveling the reverse route taken by the British convicts who first settled Australia, using the same method of transportation—hitching a ride on a ship. This species is hardy and has not only replaced native barnacles, but also oysters; it prefers the same coastal space and eats the same food as the native species.

In another instance the American oyster drill was accidentally introduced to England in 1928. The new species survived and successfully competed with the European oyster drill. But not until some particularly hard winters hit England did an imbalance occur. The American drill was not as vulnerable to cold as either the oyster or the European drill, and it became the dominant predator on an already reduced population of oysters which were dying from the severe weather.

A new species, though, may not necessarily disrupt a balance if there is an unexploited niche for it to inhabit.

In 1879 and 1882 a total of 435 striped bass were brought to the Pacific coast from their natural home in the Atlantic. Taken from the Navesink River in New Jersey, they were transported overland by train and released in San Francisco Bay. Today, they have established themselves up and down the west coast of the United States for 500 miles and they are very abundant in the Sacramento River. Their populations add to the sport-fishing catch and do not appear to interfere with the life-cycle of other fish. Such an introduction of a top predator, however, should not be made without sound research; the true impact of its presence may not be felt for some generations.

Striped bass *were introduced to the Pacific coast of the United States in the late nineteenth century and apparently have had little, if any, detrimental effect on the ecosystem.*

Predators Needed!

The word "predator" has an evil connotation, but predators are not simply destructive beasts that lurk about with bloodthirsty looks in their eyes just waiting to wreak havoc on a community. The stamina and stability of an ecosystem depend upon a solid population of carnivores.

We have seen the effects of the introduction of a new species; now let us consider what happens in an ecosystem when a species is eliminated. Robert Paine, in a classic study on predator elimination, demonstrated that removal of one species of predatory starfish from a rocky coast was detrimental to the entire system. A stretch of California shoreline, eight meters long and two meters wide, was kept free of the starfish *Pisaster*, for three years. An adjacent area of the coast was left undisturbed as a control.

Normally the coast under study has a definite zone in which two species of barnacles and the California mussel congregate across the intertidal mark. At low-tide levels a third barnacle, *Balanus glandula*, is scattered in clumps among a few anemones, limpets, some chitons, and four types of algae. A sponge and a nudibranch also occupy the site. Within four months of the removal of the starfish, the low-tidal zone barnacle, *B. glandula,* had moved into 80 percent of the experimental area. Soon this barnacle was being crowded out by young, rapidly growing mussels and the gooseneck barnacle, *Mitella*. Eventually the area was dominated by the mussel. All but one species of algae disappeared. The limpets moved out because of crowding and lack of food. The number of species in the area was reduced from 15 to 8.

By eating the shellfish, the starfish had kept living space open for a variety of animals. Once the starfish was removed, the most efficient space-grabber, the mussel, took over.

The ecosystem was greatly reduced by the artificial removal of the starfish. Instead of the proliferation of species that one would expect when the major ecosystem predator was eliminated, space competition between the prey species reduced population diversity and thus made the ecosystem, in theory, more susceptible to danger.

Predators, like starfish, *play an integral role in maintaining stability of an ecosystem; they are important biological population controls that assure variability in the system.*

Chapter V. Sea Level and Civilization

Many of man's earliest settlements are now at the bottom of the sea. The seas are rising, and one day many of the great cities that we know may be completely inundated. Since 1930 a continuous rise in sea level has been recorded around North America. The mouths of many rivers, like the Hudson and the Susquehanna, have been engulfed. The rise in sea level is greater on the Atlantic shoreline than the Pacific, but it is everywhere. We live during a period of melting of the polar ice caps, and they are giving back to the sea the water they took from it. When and where the present advance of the sea will end no one knows. There is more than enough water locked up in ice to cause a rise of 300 feet. If that happened, most of the Atlantic seaboard and the coastal plain of the Gulf of Mexico would be submerged. If the seas were to rise as much as 600 feet, large parts of the eastern half of North America would disappear under water. The Gulf of Mexico would reach the Great Lakes and the Appalachian mountain range would be reduced to a series of islands to the south and north and one 600-mile-long island in the area from Georgia to Pennsylvania.

"Like a tide too vast to measure, the seas are rising, encroaching upon the land. When and where the ocean waters will halt their advance no one knows."

The melting of the polar caps is the main reason for rising sea levels, but there are other causes. Movements of the crust of the earth are one. These movements affect the relative level of the sea at the continental margins. Each downward movement of the crust is accompanied by a movement of the sea onto the land. Each upward buckling is marked by a retreat of the sea. The displacement of ocean water by sediments washed from the land causes some of the change. So does the growth of volcanic islands.

Changes in sea level brought about by these several factors have influenced the migrations of peoples. During ice ages much of the region now known as Malaysia consisted of islands separated from one another by only shallow seas and periodically united with Asia itself. Some of the earliest human wanderers into Oceania found more exposed dry land and shallower water than exist at present. They still had many gulfs and channels to traverse, but the situation simplified their going from place to place. The migration of plants and animals from west to east can be seen in the fact that there is a far greater diversity of species in the islands closest to Asia and Australia.

The water that the ice locked up allowed North and South America to be joined at the Isthmus of Panama, and it also made it possible for man to walk across a land bridge joining Asia and North America at what is now the Bering Strait. Between India and Ceylon, too, a long outcropping bank became a shoal and finally a link between the island and the subcontinent.

All this happened during the time of the sea's retreat. Since then it has risen again and covered many of the places where man used to dwell. It is still going on, like a great tide that is too vast for us to measure.

A diver peers into ***an underwater cave.*** *Once, in an ice age, when waters froze and sea levels fell, this cave may have been a home for primitive man. Now, in an age of rising seas, it is again submerged.*

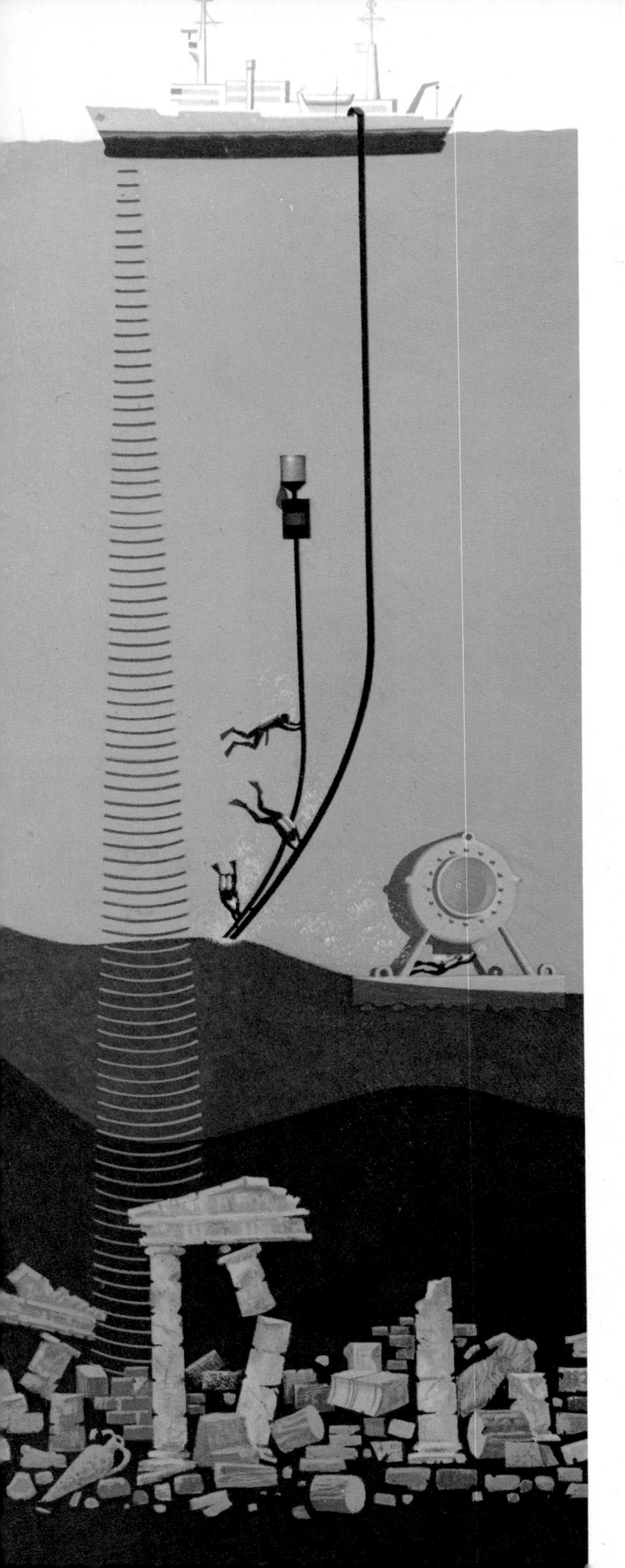

Pompeii of the Aegean

In 373 B.C. the earth's crust buckled and the entire Greek city of Helice sank beneath the Aegean Sea. Pausanius described what happened. He wrote of Helice that "it was once an important city, and the Ionians had there the most holy temple of Poseidon of Helice. The worship of Poseidon of Helice still remained with them, both when they were driven out by the Achaeans to Athens, and when they afterwards went from Athens to the maritime parts of Asia Minor.

"And later on the Achaeans here, who drove some suppliants from the temple and slew them, met with quick vengeance from Poseidon, for an earthquake coming over the place rapidly overthrew all the buildings, and made the very site of the city difficult for posterity to find.

"And they say another misfortune happened to the place in the winter at the same time. The sea encroached over much of the district and quite flooded Helice with water; and the grove of Poseidon was so submerged that the tops of the trees alone were visible. And so the god suddenly sending the earthquake, and the sea encroaching simultaneously, the inundation swept away Helice and its population."

Archaeologists yearn to salvage Helice's ancient works of art. But earthquakes frequent the area, and sonar records show that the site dropped 30 feet in the earthquake of 1870. Also, two rivers that empty into the bay bring tons of mud to bury the city more. In 1973 Harold Edgerton of the Massachusetts Institute of Technology and Spyridon Marinatos of the Greek Department of Antiquity tried to locate the ruin with sonar and drilling equipment. Under 150 feet of water and layers of mud more than six feet deep, their drill struck something hard. It could have been the temple of Poseidon.

Port of the Phoenicians

On the Mediterranean coast of Lebanon there is an insignificant little town called Sur, which means "rock." No one who visits the town today would guess that it was once one of the most important cities in the world. It was known as Tyre, and it was the principal port of the greatest traders by sea of the ancient world—the Phoenicians.

The origins of Tyre are lost in antiquity, but it is known that the port was in existence prior to 1400 B.C. At this time, a dockyard system based on a pair of offshore islands already made it an important harbor. Around 950 B.C. the two islands were connected by a mole and fortified.

Tyre and the Phoenicians prospered until 333 B.C. when Alexander the Great blockaded the city and forced its submission. To gain his objective, he destroyed the mainland town, called Old Tyre, and with its rubble built a causeway to the island fortress, making possible its capture. Tyre was later to regain a measure of its former prosperity under the Seleucids and the Romans; it then gradually declined.

From 1935 to 1937 an archaeological expedition led by a French Jesuit named Père Antoine Poidebard made a detailed investigation of the remnants of Tyre's now-submerged harbor system. When the survey began, the precise location of the ancient docks and anchorages was unknown. Much of the mapping of the site was accomplished through aerial photography. The photographs revealed the remains of the harborworks in shallow water. Underwater photography made its first appearance in marine archaeology here as well. Divers took pictures of the site; Poidebard himself took stereophotographs through a glass-bottom bucket, and he was thus able to study the ancient walls in three dimensions.

A line of submerged reefs was found to have been artificially joined and heightened to form a massive breakwater that gave protection to an ancient anchorage. Over the centuries much of it had been swept away by waves and swell, but at one end of the reef there were discovered foundation walls made of massive stone blocks, placed in such a way as to deflect silt-laden currents. The team also found two enclosed basins, complete with quays. They found a narrow channel that could be closed by a chain or other barrier and thus deny entrance to intruders.

Père Poidebard's work, including his innovative use of both aerial and underwater photography, added considerably to our archaeological knowledge of the ancient world. We now know what a Roman colonial harbor was like in the second century A.D.

Two of the most valuable ***underwater archaeological sites*** *in the Mediterranean are located at Helice in Greece, which sank beneath the sea during an earthquake in 373 B.C., and at Tyre in ancient Phoenicia, whose harbor is now submerged (see map below). The illustration at left shows how echo sounding may be used by a surface ship to determine the site of Helice and how divers might excavate, with the aid of huge air pumps, the mud that covers the ancient Greek city.*

Wickedest City in the World

British emigrants built Port Royal. They discovered the island on the southern coast of Jamaica and filled in the intervening marshland and established a settlement there. It was situated at the end of a long, low, curving sandspit known as the Palisandoes. The town was to become the headquarters of pirates as well as the crossroads of trade between the Old World and the New. The harbor at Port Royal could easily accommodate more than 500 ships at one time and its strategic position in the Caribbean made it an ideal spot from which to attack Spanish fleets carrying the silver, gold, and precious stones of the New World home to Europe. To many it was the wickedest city in the world.

Brick houses crowded the streets, spreading even to the edge of the sea. Those who built there gave no consideration to the precariousness of tall buildings resting on loose gravel. From time to time there had been a few minor earth tremors, but no damage of any consequence had been done, and people gave it little thought. Then, on June 7, 1692, "the fairest town of all the English plantations, the best emporium and mart of this part of the world, exceeding in its riches, plentiful of all good things, was shaken and shattered to pieces . . . and covered, for the greater part by the sea."

***Archaeologists** (above) work to excavate Port Royal, perhaps the most important marine archaeological site in the Western Hemisphere.*

*In the illustration at left, **buildings slide into the sea** as the city of Port Royal is devastated by an earthquake on June 7, 1692.*

***The excavation of Port Royal** has increased our knowledge of the past. Although only 5 percent of the site has been uncovered, valuable objects have been found. Pewter and earthenware dishes (top right), a pocket watch (middle right) whose brass movement was in perfect working condition, and coins and a brass keyhole plate (bottom right) are among the artifacts so far uncovered.*

The earthquake lasted a scant two minutes. Two thousand people died and two-thirds of the city slid beneath the water. Soon it was covered by mud from nearby rivers. An eyewitness gave a horrifying account: "It was a sad sight to see all that Harbour . . . covered with the dead bodies of People of all conditions, floating up and down without burial. . . . From *St. Anns* there was news that above a thousand acres of Woodland were turned into the sea . . . but no place suffered like *Port Royal,* where whole streets were swallowed up by the opening of the Earth, and Houses and Inhabitants went down together. Some of them were driven up again by the Sea . . . and wonderfully escaped."

Other contemporary accounts tell of a great seismic wave that brought ruin to everything in its path. When the flood waters receded, all that was left was a small portion of the town, and once again it was an island.

In 1959 Edwin Link did preliminary archaeological work on the site, discovering some interesting artifacts. Then, in 1965, a large-scale excavation was begun under the direction of Robert F. Marx, working for the Jamaican government. It was to become the largest marine excavation project in scope and duration ever undertaken. Marx and a few assistants worked for two and a half years and in that time were able to excavate less than 5 percent of the site.

In an average day they brought up about 300 pounds of coral-encrusted artifacts of all kinds. Among the most exciting finds were two buildings standing intact and the wrecks of two ships. One vessel was a French warship, destroyed in a hurricane in 1722. The other was H.M.S. *Swan,* which was being careened at the time of the disaster and so was without her ballast, making her light enough to be tossed into the middle of the town by a tidal wave.

Venice's Piazza San Marco *(above and opposite) is frequently flooded as a relentlessly rising sea lays claim to the most romantic city in the world.*

The prows of her ***gondolas*** *reflect the pride of the city of Venice (below) as her walls, imbued with salt and moisture from the sea, slowly crumble.*

The Indomitable Bride

For many, many years the Doge of Venice, amid great pomp and ceremony, tossed a consecrated ring into the Adriatic, saying "We wed thee, O sea, in token of our true and perpetual dominion." That dominion has ended. The Queen of the Adriatic, the city that has been called "man's most beautiful artifact," is sinking into the sea.

The early Venetians built the city that was to become the home of Titian and Tintoretto, Bellini and Giorgione, on 118 islands. They were seeking refuge from invading hordes, and the lagoon became their fortress. Fearful that it might fill with silt and so make them accessible to their enemies, they diverted three rivers that flowed into it, and they began the never-to-be-completed task of filling, bridging, and buttressing their islands.

Venice came to control the trade of the eastern Mediterranean and to rule much of northern Italy. But when the Portuguese rounded the Cape of Good Hope, opening up a new passage to India, her dominance of trade began to dwindle. The Republic steadily declined until 1797 when the last doge surrendered to Napoleon.

Today Venice is probably the most romantic city in the world. The Piazza San Marco has been called "the world's largest drawing room." Gondolas still ply the many canals that lace the city. Its great palaces are still there. But now the Piazza San Marco is frequently flooded, the houses along the canals have higher entrances and, slowly but surely, the palaces are crumbling.

The city was built just out of reach of the highest tides, one to four feet above mean sea level. Now those tides are higher, and they surge over Venice regularly—30 times since 1960. Floods have become a way of life. It is quite possible that within 150 years the sea will claim the city.

The northern Adriatic region has subsided 70 to 120 inches since Roman times. At the same time, the world's oceans have risen as the polar caps have melted. Venice itself sinks an inch every five years. It has been discovered that there exists a free oscillation of the waters of the Adriatic. This movement of the waters is superimposed on the regular tides and adds to their height. Combining with this is the sirocco, a fierce gale that sweeps up the Adriatic every now and then, sometimes blowing steadily for days, buffeting the shallow sea and driving it to surging levels.

Industry has compounded the problem. It has had a great thirst for water, and huge pumps buried in the ground have been drawing the water table down between 30 and 50 feet a year for many years. The substrata of sediments under the city has bowed and the ground subsided. Industry has also filled many of the mud flats, diked great areas for fish ponds, and cut deep channels, and all of these projects have encouraged the sea to come in.

When one walks in the church of Santa Maria Assunta dei Gesuiti now, he hears a great hollow sound. The waters have eaten an enormous hole beneath the building. The salt and the moisture remain in the walls of this and hundreds of other buildings, and some day they may crumble. The whole world is concerned about Venice, and many solutions to her problems have been offered. In June 1973 UNESCO opened an office in Venice from which it coordinates international efforts to save the city. UNESCO is providing technical assistance, scientific research, and social and cultural promotion. A mathematical model of the lagoon has been built, based on sonar-drawn charts, to determine the effects of changes in its hydrology. The model will help to predict Venice's chances for survival.

Apollonia

Apollonia, the port for the once-proud city of Cyrene on the coast of Libya, is today half submerged. Eroded marble columns embedded in sand mark its remains on land. Under Roman domination in 96 B.C., Apollonia was one of the chief export centers of the North African coast. It became fabulously rich under the Romans by exporting enormous quantities of grain.

Today Cyrene has only one street, bounded by eucalyptus trees and twisting between the low, white, Libyan shops. All around the village lie the magnificent ruins of old temples, market places, baths, and theaters. Beyond them lie the scattered remains of the suburbs, now half buried in sand and covered with thorn bushes.

At Apollonia itself there is a small fishing village near the ancient city, and some of the poor live among its dark tombs and crumbled ruins. More than half the city is beneath water. In rough weather the waves surge over the outermost walls and across the ruined harbor, tumble over the submerged buildings, and finally collapse on the sandy beach.

Apollonia has been a fertile site for archaeological research, and much information about the ancient city has been uncovered. It has been found to have been constructed on a Phoenician plan as a two-harbor city. The outer harbor received ships of commerce, while the inner harbor was fortified to protect the city from attack. An enemy ship attempting to enter the inner harbor had to pass through the connecting channel, which was defended by soldiers standing on the walls. At the end of the channel a spiked boom or chains were hung between the towers to bar the way.

Just inside one of the breakwaters is the ruin of a large submerged fish tank, divided into a number of compartments and connected to the open sea by narrow channels. In such tanks the Romans practiced ancient mariculture. The fish tank enabled archaeologists to estimate the subsidence of the city since it was built more than 2500 years ago. It had sunk six and a half feet.

One of the most startling finds at Apollonia was an anchor. It is a wedge-shaped block of stone with a hole in it about three inches across. At the opposite end are two holes perpendicular to the first. This kind of anchor is one of the earliest types known. In the *Odyssey* there is mention of the Mycenaeans using heavy stones to anchor their ships, and the first development was a stone with a hole in it for a rope. Pieces of spiked wood were probably fitted through the holes. They dug into the sea bottom and prevented the anchor from slipping.

Apollonia *(opposite). The light-colored sea area is the sandy bottom of the ancient inner harbor, and beneath the dark sea around this, the sea floor is covered in ruins. When the sea level was lower relative to the land, the harbor basin was well protected, and the islands were used for fortifications, storage, slipways, and a lighthouse.*

The submerged remains of a large artificial foundation at ***Halicarnassos*** *in Turkey (below). This huge structure is now silted up and built upon by the modern city.*

Chapter VI. Men Against the Sea

There have been shipwrecks for almost as long as there have been ships. The bottom of the sea is strewn with them and of countless thousands there is no trace. For so long as they have lain there, the sea has slowly absorbed their remains.

Great storms at sea have spread disaster and have been feared by sailors since ancient times. The ocean's waves and breakers have crushed small boats and capsized larger ones.

"The bottom of the sea is strewn with the remains of many wrecks. Many more probably will join them. Man will never cease to challenge the sea in ships and, from time to time, the sea will win."

But these are only some of its dangers. The oceans have their coral reefs and icebergs that have ripped holes in hulls to sink ships. Some vessels have sailed into the arctic and antarctic and have been crushed between great masses of ice and snow.

Some ships were lost because of man's own foolishness. Others foundered simply because they were poorly manned. Some were never seaworthy. Convoys from the Spanish "Silver Fleet" sank because they carried more gold than they could safely handle.

Many ships have been sent to the bottom by enemies in war. They were rammed with the bow of a ship or shot to pieces with cannonballs or shells. Some were destroyed with torpedoes and some with mines or depth charges. Many of them were not warships, but they carried the goods of war. A few were passenger vessels, and they carried a great many women and children.

For a long, long time man has tried to equip his ships so that he might save himself in a disaster. Naval engineers still try to develop unsinkable hulls, and in the meantime, life boats, rafts, and inflatable crafts have been the stage for scenes of great heroism as well as ruthless cruelty. Some men and women cast away at sea have shown almost unbelievable courage and a will to live that has brought them through trials that would have seemed unendurable. But in the same circumstances, others have behaved like beasts and preyed on one another.

Many men and women, most of them courageous, some of them foolish, have voluntarily set themselves the task of finding out what it is like to challenge the rigors of life at sea with very little to make their journey easy for themselves. A few of them have been determined to prove various theories about the migrations of ancient peoples or about man's ability to derive his sustenance solely from the sea. They have crossed oceans in rowboats and rafts. They have gone all around the world in catamarans and sloops. Most of those people learned some very important things about the sea and about human nature that they could never have learned in any other way.

Today, almost a million tons of new wrecks settle every year on the bottom of the sea, which is littered with the remains of those that have rotted beyond recognition. Great mounds of silt hide many of them and help preserve them. More will join them, for man will never cease to challenge the sea in ships and, from time to time, the sea will win.

***Survivors** cling tenaciously to a partially inflated life raft as their boat begins its journey to oblivion. Were wind and waves to blame, or was it human carelessness and ignorance?*

Never to Sail Again

Throughout time man has witnessed the fury and destructive capabilities of the sea. They have combined with human error to send many a ship to the bottom.

Fire rages *on a French passenger liner,* L'Atlantique *(opposite). Listing to port, it will soon go down.*

The ***drama and horror*** *of a sinking ship (right) is seen in this painting by A. Dawant dated 1889.*

Heavy seas *sent the schooner* Marjory Brown *(below) to the bottom off the coast of New Jersey.*

In 1889, a ***hurricane*** *reduced the German cruiser* Adler *(bottom) to a pile of driftwood on the beach.*

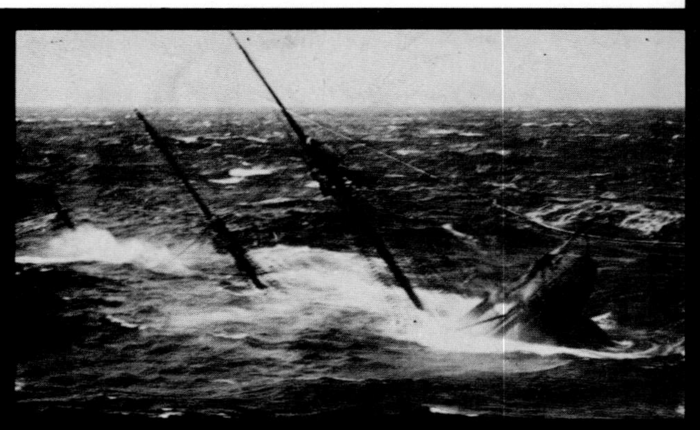

The Raft of the "Medusa," *by Jean Louis André Théodore Géricault, a French painter who lived from 1791-1824, was suggested by one of the most grisly tales in the annals of shipwrecks.*

Voyage of Horror

The wreck of the *Medusa,* with its grisly aftermath, was one of the most horrible—if not the most horrible—shipwrecks in history. On July 2, 1816, the 40-gun French frigate ran aground on a shoal about 60 miles off the west coast of Africa. The ship was not badly damaged and might have been saved if there had been courage or leadership among her crew. The day after running aground, the ship's great rudder, sweeping to and fro, smashed the stern to splinters. Water leaked in. The decks bulged. The keel was broken into matchwood. The *Medusa* had to be abandoned.

The six lifeboats could hold only about 250 of her company of 400, and so a great raft was built to hold 200 persons and provisions. The lifeboats were to tow the raft to shore. The convoy was scarcely two leagues from the *Medusa* when the towline either broke or was cut. The lifeboats made off and those on the raft were left to their fate.

One hundred and forty-six persons—one a woman—stared out on a sunlit sea. Their situation was hopeless. They had no mast, no anchor, no cable, no lines, no chart. They had canvas but no way to spread it. They were completely at the mercy of the sea. They did have a 25-pound bag of biscuits soaked with salt water, a few barrels of wine, and several casks of water. But that was about all. The wreck of the *Medusa* was still in sight, but there was no way to reach it—the wind was off the sea.

On their first night some of the men were swept away and drowned. Some were even less fortunate—they slipped between the openings as the spars of the raft opened and closed like an accordion, and they were pinned until they drowned. On the second night the survivors opened a keg and drank its wine. Madness followed. Crazed by suffering, the men started a frenzied revolt against the handful of officers. A survivor wrote that one of the officers "was seized by the soldiers, who threw him into the sea; but we perceived it, saved him, and placed him on a barrel, from which he was taken by mutineers who were going to cut out his eyes with a penknife. Exasperated by so many cruelties we no longer restrained ourselves, but charged them furiously." Many died beneath saber blows in the darkness—60 in all. One had been hacked to death with an ax, some had been held underwater until they were dead. At dawn 67 remained.

Sharks nosed about the raft now as it drove back and forth before the wind. The water was gone and so was most of the food. Some gnawed at the leather of their sword belts, and some who had hats tore off the greasy sweat bands and chewed them. It was not long before one of them began to hack away at a dead body. A moment later, dozens of them fell upon the corpse like a pack of wolves. One who lived to tell about it wrote: "Seeing that this horrid nourishment had given strength to those who had made use of it, it was proposed to dry it in order to render it a little less disgusting." In the night 12 more died. There were 48 left on the fourth day. A school of flyingfish was caught between the spars, but not enough. That night mutiny raged again on the foam-covered raft. Both sides fought desperately, and in the morning there were only 30 survivors left, all of them wounded. Most of them were unable to stand. The sea had stripped the skin from their feet and legs and now the salt water made the pain unbearable.

On the sixth day a consultation was held among the healthier survivors. They decided to throw the dying to the sea and the sharks. The woman was among them. It allowed those remaining to sustain themselves for six additional days. That sustenance consisted of 30 cloves of garlic, a small lemon, and a very small quantity of wine.

The 15 that were left drifted aimlessly on with the waves. And then, on the ninth day since they had left the *Medusa,* a white butterfly "of a kind so common in France" settled on the raft. The butterfly gave them hope and, indeed, four days later they were found and taken ashore. In the short space of 13 days they "had seen and taken part in such horrors as happily fall seldom to the lot of man." The horror was over, but it would never leave them, and it would remain in the annals of the sea as one of the most dreadful voyages ever made.

"All the News That's Fit to Print."

The New York Times.

THE WEATHER.

NEW YORK, TUESDAY, APRIL 16, 1912 TWENTY FOUR PAGES,

ONE CENT

TITANIC SINKS FOUR HOURS AFTER HITTING ICEBERG;
866 RESCUED BY CARPATHIA, PROBABLY 1250 PERISH;
ISMAY SAFE, MRS. ASTOR MAYBE, NOTED NAMES MISSING

Col. Astor and Bride, Isidor Straus and Wife, and Maj. Butt Aboard.

"RULE OF SEA" FOLLOWED

Women and Children Put Over in Lifeboats and Are Supposed to be Safe on Carpathia.

PICKED UP AFTER 8 HOURS

Vincent Astor Calls at White Star Office for News of His Father and Leaves Weeping.

FRANKLIN HOPEFUL ALL DAY

Manager of the Line Insisted Titanic Was Unsinkable Even After She Had Gone Down.

HEAD OF THE LINE ABOARD

J. Bruce Ismay Making First Trip on Gigantic Ship That Was to Surpass All Others.

The admission that the Titanic, the biggest steamship in the world, had been sunk by an iceberg and had gone to the bottom of the Atlantic, probably carrying more than 1,400 of her passengers and crew with her, was made at the White Star Line offices, 9 Broadway, at 8:20 o'clock last night.

The Lost Titanic Being Towed Out of Belfast Harbor.

Biggest Liner Plunges to the Bottom at 2:20 A. M.

RESCUERS THERE TOO LATE

Except to Pick Up the Few Hundreds Who Took to the Lifeboats.

WOMEN AND CHILDREN FIRST

Cunarder Carpathia Rushing to New York with the Survivors.

SEA SEARCH FOR OTHERS

The California Stands By on Chance of Picking Up Other Boats or Rafts.

OLYMPIC SENDS THE NEWS

Only Ship to Flash Wireless Messages to Shore After the Disaster.

LATER REPORT SAVES 866.

BOSTON, April 15.—A wireless message picked up late to-night, relayed from the Olympic, says that the Carpathia is on her way to New York with 866 passengers from the steamer Titanic aboard.

"Iceberg! Right Ahead!"

The *Titanic* was once the largest ship the world had ever known. She was also believed to be the safest. Her builder had given her double bottoms and had divided her hull into 16 watertight compartments. The world's largest ship was thought to be unsinkable.

When she set out on her maiden voyage from Southampton to New York in 1912, there were over 2000 men, women, and children aboard. They included some very prominent persons. At 9 A.M. Sunday, the third day out, a message was received in the wireless shack: CAPTAIN, *Titanic* — WESTBOUND STEAMERS REPORT BERGS GROWLERS AND FIELD ICE IN 42 DEGREES N. FROM 49 DEGREES TO 51 DEGREES W. 12th APRIL. It was cold, but the sun shone brilliantly. The floating hotel steamed on toward its destination.

At 1:42 P.M. the *Baltic* called the *Titanic* to warn her of ice on the steamer track. The operator sent the message up to the bridge. The officer on the bridge sent it on to the captain, who passed it on to one of the passengers, the director of the White Star Line to whom the *Titanic* belonged. He read it, stuffed it into his pocket, mentioned it to a couple of ladies, and resumed his promenade.

At 7:15 the captain asked that the message be returned to him so he could post it for the information of the officers. And the great ship, its speed unslackened, plowed on through the night. It was exactly 11:40 P.M. when it happened. The lookout in the crow's nest could not believe his eyes. But only for an instant, and then he knew it was a deadly reality. An enormous white shape floated directly in the *Titanic*'s path. Frantically he struck three bells. He grabbed the telephone and shouted into it: "Iceberg! Right ahead!"

The headlines of the **New York Times** *(left) on the morning of April 16, 1912, tell the shocking news of the* Titanic's *ill-fated maiden voyage.*

The iceberg ripped a ***300-foot hole*** *in the* Titanic's *hull, flooding six of her watertight compartments (above). She sank in less than 2½ hours.*

The Titanic ***disregarded iceberg warnings*** *and proceeded full speed ahead. When a berg appeared in the ship's path, disaster was inevitable (above).*

Sinking *about 400 miles from Newfoundland, the* Titanic *(below) took with her around 1635 people. Only 712 persons were saved.*

In the engine room the indicators on the dial faces swung around to "Stop" and then to "Full Speed Astern." But it was too late.

Curiously, the collision did not even wake many of the passengers. To others there seemed to be only a slight jar and a crunching sound. Few could have guessed that the berg had torn a 300-foot gash in the bottom of the ship and that the sea was already surging into the hold. Orders were given that the band take their places on the deck and play—morale had to be kept up. Some of the musicians were to continue to play until the ship was almost gone. Rockets were sent up in the hope that a nearby ship might see them. Lifeboats began to go over the side—only half-filled. Finally the *Titanic* stood on end and began the plunge to the depths, slowly at first, then quicker and quicker. The forward funnel snapped and killed swimmers beneath it. At 2:20 A.M., just a little less than two and a half hours after the collision, the *Titanic* was gone.

In the morning, most of the 712 who were rescued finally saw the iceberg that had sent the *Titanic* to the bottom. It was tinted with the sunrise, floating idly in the sea, pack ice jammed around its base.

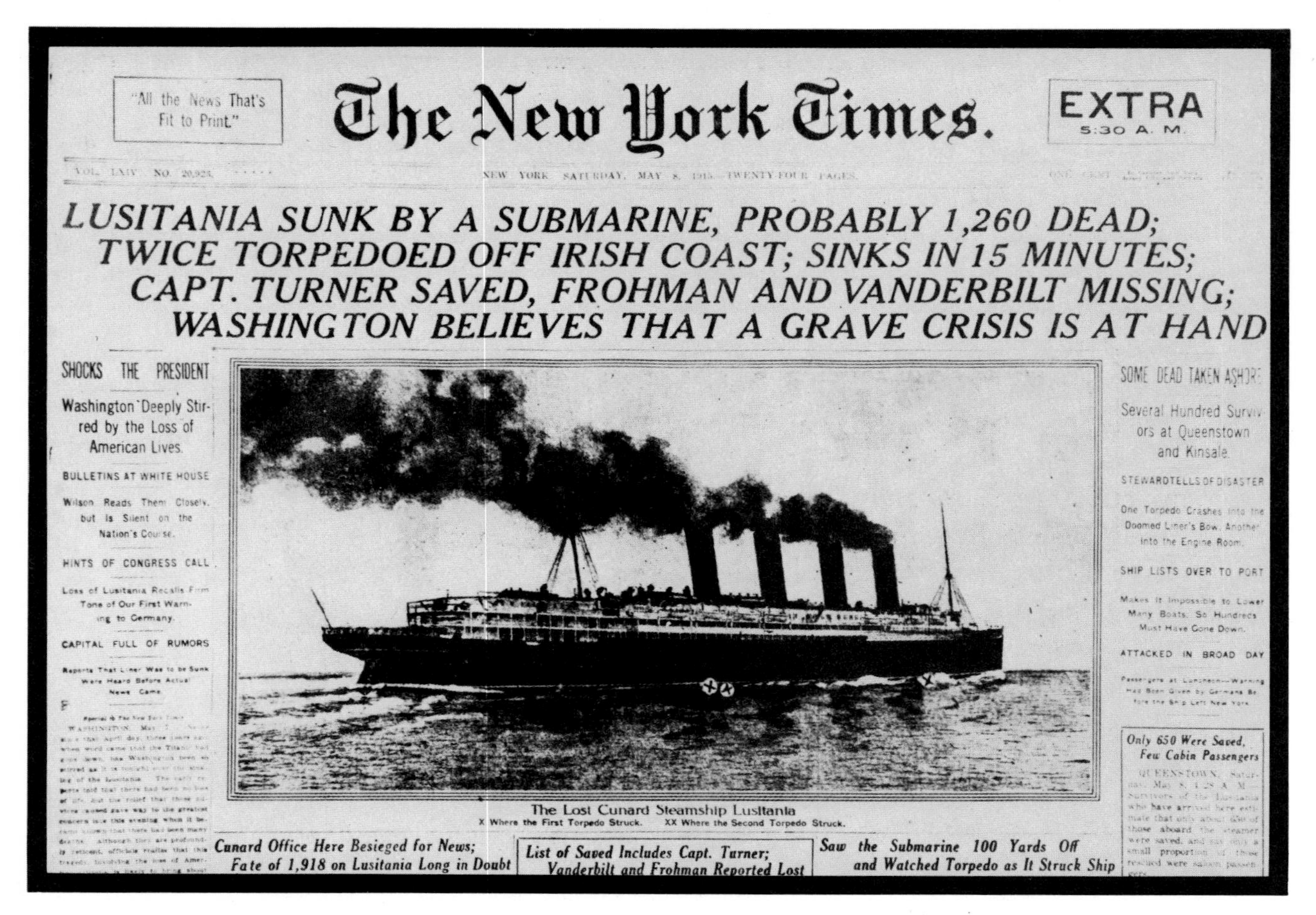
"All the News That's Fit to Print."

The New York Times.

EXTRA 5:30 A. M.

NEW YORK, SATURDAY, MAY 8, 1915.—TWENTY-FOUR PAGES.

LUSITANIA SUNK BY A SUBMARINE, PROBABLY 1,260 DEAD; TWICE TORPEDOED OFF IRISH COAST; SINKS IN 15 MINUTES; CAPT. TURNER SAVED, FROHMAN AND VANDERBILT MISSING; WASHINGTON BELIEVES THAT A GRAVE CRISIS IS AT HAND

SHOCKS THE PRESIDENT

Washington Deeply Stirred by the Loss of American Lives.

BULLETINS AT WHITE HOUSE

Wilson Reads Them Closely, but Is Silent on the Nation's Course.

HINTS OF CONGRESS CALL

Loss of Lusitania Recalls Firm Tone of Our First Warning to Germany.

CAPITAL FULL OF RUMORS

Reports That Liner Was to be Sunk Were Heard Before Actual News Came.

The Lost Cunard Steamship Lusitania
X Where the First Torpedo Struck. XX Where the Second Torpedo Struck.

SOME DEAD TAKEN ASHORE

Several Hundred Survivors at Queenstown and Kinsale.

STEWARD TELLS OF DISASTER

One Torpedo Crashes Into the Doomed Liner's Bow, Another Into the Engine Room.

SHIP LISTS OVER TO PORT

Makes It Impossible to Lower Many Boats, So Hundreds Must Have Gone Down.

ATTACKED IN BROAD DAY

Passengers at Luncheon—Warning Had Been Given by Germans Before the Ship Left New York.

Only 650 Were Saved, Few Cabin Passengers

Cunard Office Here Besieged for News; Fate of 1,918 on Lusitania Long in Doubt

List of Saved Includes Capt. Turner; Vanderbilt and Frohman Reported Lost

Saw the Submarine 100 Yards Off and Watched Torpedo as It Struck Ship

Emissaries of Death

Few people saw the advertisement in the *New York Times* on May 1, 1915. Paid for by the Imperial German Embassy, it said:

"Travellers intending to embark on the Atlantic voyage are reminded that a state of war exists between Germany and her allies and Great Britain and her allies, that the zone of war includes the waters adjacent to the British Isles; that in accordance with final notice given by the Imperial German Government vessels flying the flag of Great Britain or any of her allies are liable to destruction in those waters and that travellers sailing in the war zone on ships of Great Britain or her allies do so at their own risk."

On the very same day the Queen of the Seas, the *Lusitania,* steamed out of New York harbor bound for England. At approximately the same time the German submarine U-20, under the leadership of Lieutenant Commander Walter Schwieger, cleared Borkum Roads bound for the Irish Sea.

The 1959 passengers and crew of the *Lusitania* included a large number of Americans, and America was not in the war. For most of them it was an Atlantic crossing like any other. By evening of May 6 the German U-20, cruising in the fairway of St. George's Channel and in the steamer lanes around Ireland, had bagged three ships.

On the next day some of the passengers who were out on the deck of the *Lusitania* noticed a peculiar white feather of foam some distance from the ship and then a long streak of bubbles that came through the waves toward the side. They did not know what it was. But a sailor on the forecastle head had no doubt. He shouted through a megaphone: "Torpedo coming on starboard side!"

Commander Schwieger, aboard the U-20, had seen the four funnels and two masts of the steamer through his periscope. He would later be able to note in his diary—"clean bow shot from 700 meters range. . . . Shot hits starboard side right behind bridge. An unusually heavy detonation follows with a very strong explosion cloud. . . . She has the appearance of being about to capsize. Great confusion on board."

And there was great confusion on board. Two lifeboats were frantically filled without orders or direction and were lowered, scraping against the ship's side. Suddenly one of them tilted down at the bow, dumping its cargo of women and children into the sea. Seconds later the other boat met the same fate.

After being hit, the *Lusitania* plowed on, listing as the ocean roared in, but she did not have long to live. The stern lifted just before she went down, and her screws, still feebly turning, killed swimmers in the water. Many lifeboats that had gotten down were caught beneath the ship when she turned on her side and they went down with her.

That night fishermen and rescue ships swept the sea and brought to shore the living and some of the dead. There were 758 survivors. They spoke later of the terrible sound that rose above the sea when the ship foundered and thousands struggled for their lives. They said it resembled a "long, lingering moan."

It has long been rumored that the *Lusitania* was carrying more than just passengers. She was said to be fair game for the U-boat because her holds contained a large shipment of ammunition destined for beleaguered Britain. In recent years divers have gone down to the wreck, but no proof of explosive cargo has yet been brought back.

The **N.Y. Times** *(left) tells the tragic story. Many of the* Lusitania's *passengers were thrown from lifeboats and pulled under with the ship (below).*

The Spithead Wrecks

Sir Henry Howard had reported in a letter to his king, Henry VIII of England, that "the *Mary Rose* is your good ship, the flower I trow, of all ships that ever sailed." On Sunday, July 19, 1545, King Henry stood on the shore at Portsmouth and watched the *Mary Rose* catch a faint breeze and move out ahead of the fleet that was prepared to meet an imminent invasion by French forces.

Suddenly the *Mary Rose* heeled over and sank. No one knew for sure what caused the disaster, but an eyewitness said that in his opinion the ship had been manned by "too many knowledgeable sailors, each of whom thought he knew best, with the result that orders were not properly carried out."

When the French forces had been repulsed, operations began to try to raise the ship. A method was employed that had been invented by the Venetians, and some of them were even hired to help with the job. The method made use of pontoons. These were positioned over the sunken ship and then, by means of ropes and capstans, the wreck was to be hoisted off the bottom. The pontoon method was to be aided by a rising tide that gave a 14- or 15-foot lift. The salvagers estimated that if they hoisted the ship off the seabed at low tide, the rising tide would lift both the pontoons and the ship well clear of the bottom. Then she could be towed to a mud flat where she would be left high and dry when the tide went out.

The machinery for the hoisting off the bottom was impressive: "Two of the greatest hulks that may be gotten: four of the greatest hoys within the haven: five of the greatest cables that may be had: two great hawsers: ten capstans with twenty pullies: fifty pullies bound with iron: five dozen ballast baslets: fourty pounds of tallow: thirty

Siebe, Gorman & Company, Ltd.

Some time after the **Royal George** *sank in 1782, an attempt was made to raise her. Ropes were slung under her but she was too heavy to lift.*

Venetian mariners and one Venetian carpenter: sixty English mariners and a great quantity of cordage of all sorts." But with all this elaborate salvage equipment, the *Mary Rose* was too heavy to raise.

The ship remained one of the most famous of wrecks, and was popular with scavengers. Then, in 1968, a few concerned marine archaeologists became "tenants" of 1200 square feet of seabed at Spithead, where the remains of the *Mary Rose* still lie. By doing so, they have succeeded in protecting what is left of the wreck from plunder by commercial salvors and souvenir collectors.

In another tragic episode England was at peace when she lost her greatest warship. The year was 1782, and the 108-gun *Royal George* was host to a large number of visitors as the ship stood in Spithead Harbor. A small leak was discovered just under the waterline. The ship was listed slightly to raise the hole above water—by moving some of her guns from one side of the deck to the other. But the strain of all the additional weight on one side was too much for her, and some of her timbers cracked under it, causing the ship to list even further until the gunports went under and she quickly capsized. About 900 persons were drowned as she went down. Some time later a salvage operation was undertaken. Ropes were slung under the ship's carcass by divers in primitive diving helmets, but the *Royal George* refused to budge. This great warship was simply too heavy to lift out of her watery grave.

For many years the remains of the ship were a menace to shipping in the area. Finally, in 1839, a second attempt was made to raise her off the bottom. It was just at this time that Augustus Siebe, a brilliant German inventor, had developed a diving suit that allowed much greater freedom underwater than was possible earlier, and his equipment was called into use. Unfortunately, the divers discovered that the *Royal George* was too rotted to salvage as a whole. The ship was broken up with explosives.

A second attempt *to raise the* Royal George *was made in 1839-1842. The hull was too rotted to salvage whole and had to be broken up with explosives.*

The ***U.S.S.* Thresher** *(above left) and the* ***U.S.S.* Scorpion** *(above right) were two nuclear submarines that met with disaster at sea.*

Search for Lost Submarines

By the 1960s submarines and submarining had come to be thought of as a relatively safe business in time of peace. In the 18 years since World War II only two United States submarines had been lost—the *Cochino,* sunk off Norway in the summer of 1949 following an explosion of hydrogen gas, and the *Stickleback,* rammed and sunk by an escort ship near Hawaii in 1958. A civilian engineer aboard the *Cochino* had been killed as well as seven sailors from the rescue ship. No men were lost when the *Stickleback* sank. Thus a sense of security about submarine operations had grown in the public mind, and with about 100 U.S. submarines making a combined total of 50,000 dives each year, it was not unjustified. Understandably then, there was profound shock and sense of loss when a submarine with 129 men aboard went to a watery death off Cape Cod, Massachusetts, on April 10, 1963. To compound the shock was the fact that the nuclear-powered *Thresher* was the most advanced military submarine in existence.

We do not know for certain why the *Thresher* was lost. We know little more than the fact that at 8:53 A.M. on that fateful day the *Thresher* radioed that she was proceeding to test depth, which for *Thresher* was somewhere between 800 and 1000 feet.

At 9:12 A.M. the *Thresher* indicated in a brief message that she was experiencing some kind of difficulty. A few minutes later came a garbled message; at least one man who was present on the listening surface ship claims that it contained the words "test depth." Seconds later the men on the surface heard a sound like "a ship breaking up . . . like a compartment collapsing."

Within hours aircraft and ships reached the scene. The aircraft put their electronic and magnetic search devices into immediate operation while the ships began probing the depths with sonar. It was not long before a patch of oil and bits of debris were discovered on the surface. There was no doubt about it—the *Thresher* was lost.

The search for the *Thresher* now became one of the most extensive ever undertaken at sea. Several destroyers and more than a half-dozen survey ships, a submarine rescue ship, as well as the most advanced oceanographic research ship, the *Atlantis II,* and another research ship, the *Conrad,* arrived to join the effort. The bathyscaphe *Trieste* came and made a number of dives to the floor 8400 feet beneath *Thresher's* last reported location. Big underwater cameras with powerful strobe lights and sounding equipment were lowered to positions 15 to 30 feet from the bottom. A mosaic of the photos taken of debris in the area positively identified the wreckage as the *Thresher.*

Five years later, another nuclear submarine, the *Scorpion,* was on its way home after completing a training mission in the Mediterranean. With a crew of 99 aboard, the *Scorpion* was due to arrive in Norfolk, Virginia, in May 1968, but it never arrived. A massive hunt began, but unlike the *Thresher* searchers, no one knew where to look.

While planes, surface ships, and submarines combed the Atlantic, the U.S. Defense Department revealed that the *Scorpion's* mission was not entirely a training cruise—it had been involved in an assignment that was classified. In addition, technical reports showed that there were minor mechanical problems with some of the equipment aboard the submarine and that the emergency surfacing system was defective and the regular surfacing system was considered by some navy officials to be inadequate.

While debate about the seaworthiness of the *Scorpion* went on, the search was narrowed to the area around the Azores Islands in the North Atlantic Ocean where the sub's last radio message had been received. The ocean research vessel *Mizar* finally photographed the wreck about 450 miles southwest of the Azores and later reports said that navy underwater detecting devices had recorded an implosion of the hull. The *Mizar* photographs showed the hull was flooded in several places but that the ship's superstructure was intact, about 100 feet from the bow.

***The hull of the* Scorpion,** *photographed in over 10,000 feet of water by the research ship* Mizar.

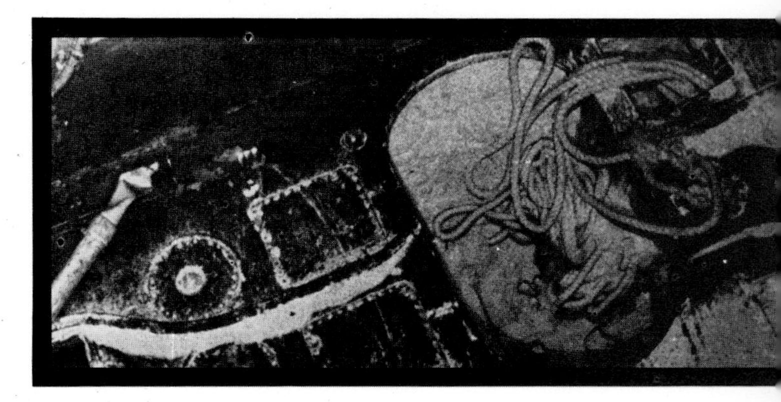

Mooring line protrudes from a ***hatchlike opening*** *in the aftersection of the* Scorpion's *hull.*

The remains of the Thresher's ***external sonar dome*** *photographed by the bathyscaphe* Trieste.

The **Johnson-Sea-Link,** *operated by the Smithsonian Institution, is loaded aboard its mother ship. In 1973 this small submersible got caught in the maze of lines and masts of a submerged wreck, and before it could be freed, two of its crewman died of carbon dioxide poisoning and cold.*

Caught in a Wreck

Off Key West, Florida, lies one of many ships of the U.S. Navy that has been deemed no longer fit for duty and scuttled. It is the hulk of a World War II destroyer, and oceanographers were interested in the possiblity of making it an artificial reef that might attract sea life for study. To investigate the question, the submersible of the Smithsonian Institution, the *Sea-Link,* went down to take what would first be a leisurely look.

The *Sea-Link* is a submersible with two separate, unconnected compartments. The pilot, Archibald Menzies, and an observer, marine biologist Robert Meek, rode in the front one, a transparent acrylic sphere. The larger rear cabin was an aluminum lockout chamber that could be pressurized. On this Sunday the two men occupying the rear chamber, Albert Stover and E. Clayton Link, had not taken along diving equipment.

One hour into their mission, they became trapped in the cables and debris surrounding the wreck, 60 fathoms down. Their SOS set off a rescue effort that lasted an excruciating 32 hours. The submarine rescue ship *Tringa* and the navy's roving diving bell, as well as the oceanographic vessel *A. B. Wood II,* joined the mother ship *Sea Diver* over the site of the accident.

Both spheres of *Sea-Link* had Baralyme systems for air purification. But aluminum conducts heat better than acrylic, and as the temperature in the rear compartment dropped from 75° F. to 40°, the absorption capacity of the Baralyme was reduced by half. After 22 hours, with the carbon dioxide content of their air mounting, the men pressurized the cabin to equal water pressure at 80 feet, then switched to helium and oxygen. Helium, with its high heat-absorbing capacity, further intensified the cold.

The first rescue divers found that they could not get near the submarine because of the maze of lines and masts that engulfed the ship. The diving bell descended, but it found the current too strong to operate effectively. It was the oceanographic vessel which, with the aid of underwater television, managed to get a grapnel on the minisub and hoist her to the surface. The two men in the forward compartment spent 90 minutes in the decompression chamber and then were pronounced well and fit. But when the rear compartment had been decompressed, both its occupants were found dead. They were the victims of carbon dioxide poisoning, anoxia, and cold, the latter being the precipitating factor. One of the victims was the son of the man who had designed the minisub, Edwin Link, who had watched the rescue operations from *Sea Diver.*

Songs from the Seabed

Not all accidents end in tragedy; sometimes luck is on the side of the submariners. About two months after the *Sea-Link* disaster, two Britons were rescued from a submersible trapped on the floor of the Atlantic about 100 miles off the coast of Ireland.

The men, Roger Mallinson and Roger Chapman, were aboard the *Pisces III* helping to lay a transatlantic telephone cable between England and Canada. Their job was to use a water jet to gouge out a trench, then nudge the cable in. At the end of a nine-hour workday on August 29, 1973, *Pisces* was being hauled to the surface by its mother ship, the *Vickers Voyager*. Capt. Leonard Edwards of the *Voyager* explained what happened. "While still on the surface, the hawser apparently tore off an atmosphere hatch, water poured into the flood compartment and the submarine sank. When it was 170 feet down, the hawser snapped and the submarine dropped to the bottom."

It was there that *Pisces* rested, mired in the sediment of the ocean floor, 1375 feet below the surface and with only a 72-hour oxygen supply. To make matters worse, gale force winds, heavy seas, and poor visibility on the sea floor hampered rescuers looking for the

20-foot-long craft. About the only thing in the trapped men's favor was that they were able to maintain radio contact with the surface. They used valuable oxygen to stay on the radio in order to guide sonar toward the craft by singing Irish sea chanties and the Beatles' song *We All Live in a Yellow Submarine*. Mallinson spent his 35th birthday in *Pisces,* literally singing for his life.

The rescue was made three and a half days after the ordeal began. *Pisces II* and *Pisces V,* sister ships of the trapped sub, and an unmanned submersible were able to guide lifelines around *Pisces III,* which had been resting at a 70° cant. When Mallinson and Chapman finally reached the surface, there was only about 90 minutes of air left in the sub.

In 1973, **Pisces III** *was trapped below 1000 feet while laying a cable. The sub and her two crewmen were rescued (above and below) after 76 hours.*

"A Victory from the Sea"

On May 15, 1939, the *Squalus,* one of the U.S. Navy's newest submarines, began a trial dive in the North Atlantic. The depth indicator moved rapidly as the submarine, whose name means "shark," dove to periscope depth . . . 35 feet . . . 40 feet . . . 45 feet . . . At 50 feet the skipper cried out, "Mark." The submarine halted her descent, and he gripped the handles of the number one periscope and peered through. As he did, he heard a strange fluttering sound. That sound was the rush of air being shoved violently forward by the ocean as it burst into the submarine. Somehow the big main air-induction valve leading to the engines had failed to close or, if it did close, had opened again. The engine rooms were flooding. "Blow all main ballast! Blow bow buoyancy! Take her up! Take her up!" he shouted. But the *Squalus* would not ascend. It shuddered and made a desperate effort, but the growing weight in her tail was too much for her. She began to sink to the bottom. Three forward compartments had been made secure. Of the 59 men on board, only 33 could be accounted for.

When the submarine was finally missed, her sister ship, the *Sculpin,* was sent to have a look. The *Sculpin* might never have found her if a young officer on her bridge had not paused for an instant to wipe the spray out of his eyes. He glanced the wrong way at precisely the right moment, for there on the water was the smudge of a smoke bomb sent up by the trapped ship.

The men alive aboard the *Squalus* were rescued, not by the *Sculpin,* but by one of the first rescue compartments—the *S-1,* which was operated by a cable from its surface ship, the *Falcon.* All 33 men were brought to the surface. In the *New York Times* the next morning the story began, "Man won a victory from the sea early this morning."

The U.S.S. **Squalus** *comes into sight (above) as men from the rescue ship* Falcon *struggle to bring the stricken submarine to the surface of the Atlantic.*

The tender Wandank *and the rescue ship* Falcon ***anchored over the sunken*** Squalus *(below). Aft on the* Falcon *is the rescue chamber that brought 33 members of the* Squalus's *crew to safety.*

Ready for Rescue

Since 1920 more than 1000 men have died in 29 U.S. submarines accidentally sunk in peacetime at depths below their hull-collapse limits. Officials of the U.S. Navy have estimated that all of these disasters offered at least the possibility of crew rescue if only there had been a rescue submarine capable of very deep dives. Today such a vessel exists.

The first of the navy's Deep Submergence Rescue Vehicles (DSRV) was launched in 1970. Measuring nearly 50 feet in length, the DSRV is designed to carry a three-man crew and have the capacity for retrieving 24 survivors at a time from a disabled submarine. Its maximum operating depth is 5000 feet, well below the collapse depth of any existing military submarine.

Upon receiving word of a submarine in distress, the nearest available pair of DSRVs is assigned to the mission. Surface support ships are dispatched to the scene, and air cargo planes or nuclear submarines transport the DSRVs to the site. At the site of the accident, the DSRVs detach themselves from the transport submarines and, using their sensors, approach the stricken submarine. Each DSRV maneuvers itself into position over either the forward or aft escape hatch, settles down on it, and latches itself securely to it. The mating section, an airlock, is pumped free of water and the hatches are opened to receive the first crewmen. Oxygen tanks are passed through to assure a safe atmosphere in the submarine during the operation. With the hatches closed and the pressure equalized, the DSRV disengages and returns to the nuclear submarine for the transfer of the survivors. Submarines of the U.S. Navy carry from 100 to 160 men. Two DSRVs can complete a rescue in two hours.

DSRV can only attach itself to standard-size hatches, such as those adopted today by all submarines of the U.S. and most of those of the members of the North Atlantic Treaty Organization. Unfortunately, none of the existing exploration submersibles has been equipped with such hatches.

The U.S. Navy's ***DSRV*** *(Deep Submergence Rescue Vehicle), seen above aboard a transport submarine and underwater, is capable of retrieving 24 survivors at a time from a disabled submarine. Nearly 50 feet long, she can descend to a depth of about 5000 feet, well below the hull-collapse depth of any existing military submarine.*

High-Risk Salvage Operation

How do you find an object twelve feet long and two feet thick in the ocean at a depth of half a mile and then pick it up and bring it home? If the object is an H-bomb, an answer must be found. It was.

In January 1966 two U.S. Air Force planes, one a bomber, plummeted to earth after one of them exploded, damaging the other plane. The accident took place six miles over Spain. The B-52 was carrying four unarmed H-bombs. Three of them fell on the ground. One of them dropped into the sea.

Twenty-five navy ships and a whole armada of deep-diving submarines gathered off Palomares, a fishing village on the southern coast of Spain, for the largest concentrated underwater search in history. The H-bomb had to be recovered. Although it was not armed and could give off no lethal radiation, it is not easy to convince people of this, and the fact that it was there struck fear into the hearts of many people. Also there was the remote possibility that a potential enemy could retrieve the bomb.

Side-looking sonar, a robot 12 feet long and weighing 1500 pounds, took part in the search. The robot was towed back and forth by minesweepers, 200 to 400 feet off the bottom, crisscrossing a 45-square-mile patch of the Mediterranean. The robot scanned the ocean floor with its beam of high-frequency sound, and spotted over 260 items on the ocean floor, most of them debris from the planes. It helped to make the final determination of where the bomb was.

It was the *Alvin,* the deep-diving submersible from Woods Hole Oceanographic Institution, that made the final discovery. On *Alvin*'s tenth dive the men aboard her noticed a groove in the undersea mountainside that could have been made by the H-bomb moving down the slope. The groove led *Alvin* to an object that was identified as a parachute of the kind attached to a bomb in order to slow its descent and give the plane a chance to escape before the explosion. The men on the *Alvin* also saw a silhouette they thought was the bomb. They photographed it at 2532 feet. The developed pictures revealed the bomb. Two months after it had fallen into the sea, it had been found.

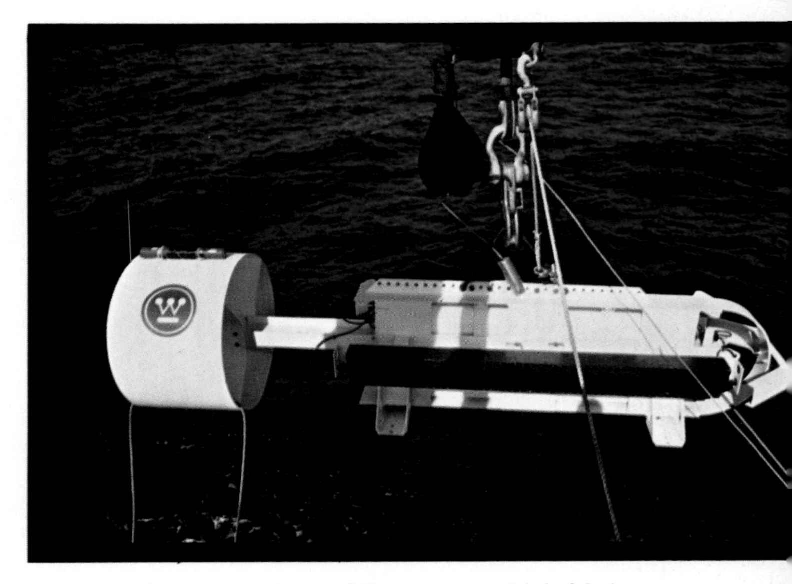

Now came the job of bringing the bomb to the surface. No lost object had ever been found and brought up from half a mile down. The *Mizar,* a navy oceanographic ship, parked right over the site. The *Aluminaut,* another deep-sea submersible, took turns with *Alvin* keeping a close watch on the bomb lest it slide into a crevice. With her mechanical arm the *Alvin* fastened a line from the *Mizar* to the bomb. Then the bomb began to be hauled upward. Suddenly the line broke, and the bomb fell to the bottom once more. It was nine days before *Alvin* found it again. It was deeper now—2850 feet to be exact. This time the site was marked with a flashing light and pingers to emit sound. Then along came CURV.

CURV is an underwater robot whose name is an acronym for Cable Controlled Under-

An ***H-bomb lost off the coast of Spain*** *in 1966 was found with the help of a robot with side-scanning sonar (opposite page). Then* Alvin *(top left), a deep-diving submersible, made the precise determination of location and another robot called CURV (bottom left) managed to connect grappling hooks to the bomb. After considerable difficulty the bomb was safely deposited on the deck of a ship (above).*

water Recovery Vehicle. It is run by cable controls from a surface ship where it is watched on television. With her mechanical arm, CURV tried to pick up the bomb. She couldn't manage it. New tools had to be invented on the spot. A four-point grappling hook was developed, and a nylon line was attached to it. CURV shoved the hook into the shrouds of the parachute. *Alvin* kept watch, almost fatally entangling herself, while CURV went up for a second hook. CURV came back and drove it into the shrouds. She tried to pin a third one into place but gave up after almost getting entangled herself. Two proved to be enough. It took just an hour for the bomb to be carefully hoisted to the ship above. As the bomb approached the surface, navy frogmen attached additional lines to it. Finally the bomb, a few barnacles already clinging to it, was deposited on deck. The greatest underwater search in history was over.

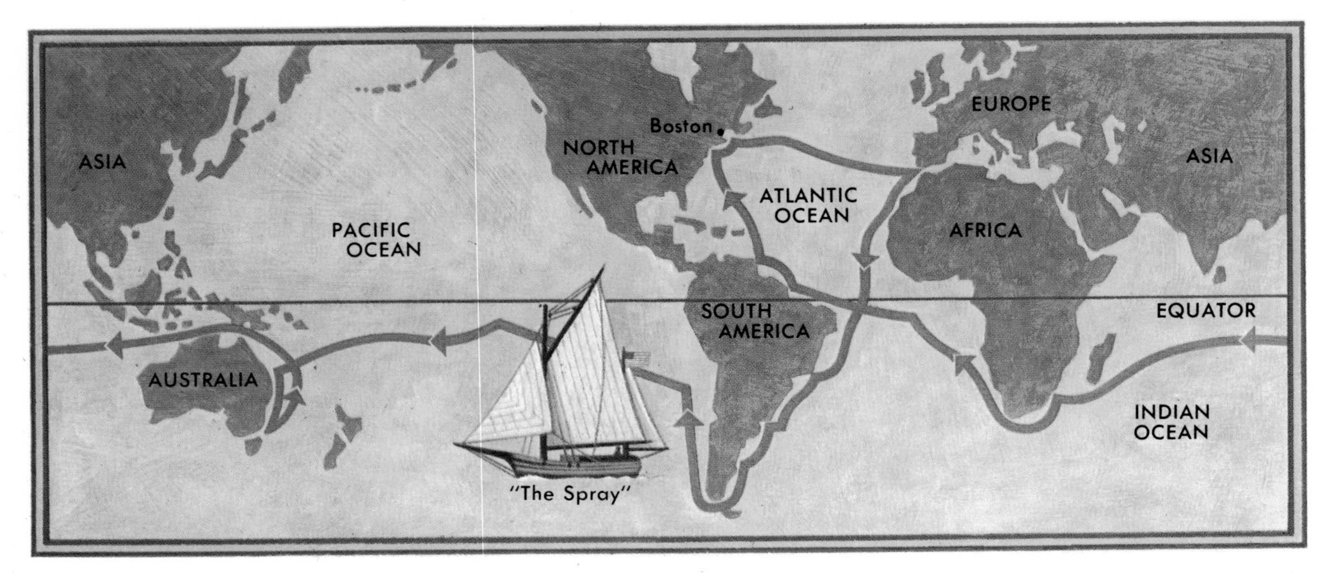

***The route** followed by Captain Joshua Slocum in his 35-foot craft covered 46,000 miles.*

Solo Circumnavigation

One of the greatest lures of the sea is the challenge of circumnavigating the globe. Every sailor knows that all it takes to accomplish this are a seaworthy craft, food and water, and knowledge about wind, water, and navigation. Yet it took until the sixteenth century before man—in the persons of Magellan's men—sailed around the world. And it was only within the last century that a lone man accomplished this feat.

The lone navigator in this case was Captain Joshua Slocum, a native of Nova Scotia who became a U.S. citizen. The trip around the world was no foolhardy venture, for Slocum had considerable experience piloting fishing craft and cargo vessels. He planned his journey well, even going so far as to build his own boat, the 35-foot-long *Spray,* using the wreck of a 100-year-old oyster boat as a base to build on. Slocum shaped his vessel with new wood, patience, hard work, and an unheard-of caulking made of cotton and oakum.

At the age of 51, this daredevil mariner set sail from Boston on April 24, 1895, and headed across the Atlantic, past the Azores toward Gibraltar, where he received a tumultuous welcome. He then pointed southwestward, crossing the Atlantic again before putting in first at Pernambuco and then at Rio de Janiero, Brazil. He worked his way along the South American coast and reached the Rio de la Plata between Uruguay and Argentina by Christmas.

The trip was not without many potential hazards. Once, Slocum reported, a whale made his presence known by scratching his back on the keel of the *Spray.* But rather than frightening the captain into turning back, Slocum became enchanted with leviathans and wrote, "for broad, rippling humor the whale has no equal."

During these first months of the journey, Slocum discovered the problems that beset solitary navigators, such as the immense loneliness that even his store of books couldn't always alleviate. And he quickly realized, after eating foul food and jettisoning his supply of plums, that if he were ill, there was no one to man the ship. Slocum persisted and entered one of the roughest parts of his trip—through the Strait of Magellan—during the height of summer in the Southern Hemisphere. The summer weather, perhaps, was the only reason he was able to make it, for he was still buffeted by heavy seas and high winds.

The loneliest part of the journey still lay ahead, though, for the southeastern Pacific is notoriously devoid of islands, and, as the captain noted at the end of his log, the *Spray* did not discover any new continents. He pushed on to the Marquesa Islands, Samoa, Fiji, and New Caledonia enroute to Australia, where he spent Christmas of 1896. When he resumed his travels, he went north toward New Guinea before heading west again, transversing the Indian Ocean past what is now Indonesia, but known to Slocum as the Dutch East Indies.

One danger faced by Slocum came not from wind or water, but from land, since he was never certain what kind of reception he would receive when he put ashore, especially for unplanned stops to make repairs. Slocum was a sailor, not a soldier, and he could easily have run into some unfriendly natives and never been heard from again.

Heading toward Africa, the *Spray* passed the strange island world of Madagascar and reached the Cape of Good Hope in enough time for Slocum to spend Christmas of 1897 in Capetown. The homeward leg was northwesterly across the Atlantic, touching Ascension Island and then Antigua in the Caribbean before heading up along the coast of the United States to return to Boston on June 27, 1898.

The intrepid Slocum had covered more than 46,000 miles on his journey and was none the worse for wear, as he explained it. "I had profited in many ways," he later wrote. "I had even gained flesh and actually weighed a pound more than when I sailed from Boston... And so, I was at least 10 years younger than the day I felled the first tree for the construction of the *Spray*."

His conclusion was not one of fear and loathing of the dangers he had encountered and the harrowing experiences he had been through. Rather, he wrote, "The sea has been much maligned . . . and the *Spray* made the discovery that even the worst sea is not so terrible to a well-appointed ship."

The **Spray** *was built by Captain Slocum himself, based on the wreck of a 100-year-old oyster boat.*

Surviving at Sea

Long ocean voyages may challenge the skilled navigator in his well-appointed ship, but the same trip is sheer hell to a survivor of a plane wreck. This kind of experience was not uncommon during World War II with its aerial warfare. Many pilots, bombers, gunners, and radiomen had to take to the water after their planes were shot down or otherwise disabled. One of the most harrowing of these ditchings involved a three-man crew that crash-landed in the South Pacific in January 1942, during the height of the war. Pilot Harold Dixon, bomber Tony Pastula, and gunner Gene Aldrich were soon drifting in a GI survival raft that measured four feet by eight feet on the outside. On the inside though, the rubber-and-fabric craft had a space of only 40 inches by 80 inches for the three men. They had to stay alive, avoid the enemy, and find land; and their only tools were a pocket knife, a pair of pliers, and a 15-foot length of line. There had been no time to grab anything else.

The knife was the most useful of the objects, for it could be used to stab at the curious tropical fish that were attracted to the yellow-orange survival raft. And there were some birds in the area, mostly albatrosses and petrels, curious to see what manner of creatures these men were. The most successful kill, though, was a shark, which provided the only good meal of the ordeal.

The sun's heat was also a problem, because it blistered their skin. Even dousing clothing with seawater and wearing it as protective head coverings provided only temporary relief. The cool, clammy night air was only slightly more bearable than the sun-baked days. Freshwater was provided by the infrequent rainfalls and was soaked up by the rags of clothing.

Dixon, Aldrich, and Pastula maintained enough sanity not to do anything foolish, although they were half-crazed with thirst and starvation and weakened with despair. They drifted for 34 days and traveled about 1000 miles without being spotted by a plane or submarine, American or Japanese. They finally drifted onto a coral atoll that was inhabited by friendly natives and, luckily, a British commissioner who radioed for help.

I witnessed one of the oddest battles of man against the sea. We had *Calypso* in antarctic waters during the austral summer of December and January of 1972–73 and had been

alerted to be on the lookout for the *Ice Bird,* a small vessel that was in distress. The only person aboard was a David Lewis, a medical doctor from Australia. The chances against spotting another ship on the high seas are as great as finding the proverbial needle in a haystack, perhaps greater.

On the morning of January 29, I awoke about 4 A.M.—it was broad daylight then—to find a small boat alongside the *Calypso.* This was quite a surprise in itself, and I assumed the boat was abandoned and somehow drifted into our ship. I hurried topside—for we didn't want some wreck to damage the *Calypso*—where we were startled to see a man appear. His hair and beard were messy, his face reflected his ordeal. His spirits were good, however, despite the fact that his clothes were still wet from the time two months earlier that he had capsized, losing his mast. Shortly before, he had set out from Hobart, Tasmania, to circumnavigate Antarctica. He had managed to proceed 2500 miles with a makeshift mast, but was without radio or heater and was forced to eat only cold tinned food and biscuits since everything else had been wet or lost.

Dr. Lewis came aboard *Calypso* and told his story. He asked us to relay a message to his two daughters, aged 10 and 12, back home. It read: "Sorry, the boat is a bit broken, but I will mend it here. Hope you were not too worried. I miss you both so much. Was Christmas nice? I shared mine with the little ice birds. They had broken biscuits. There are thousands of penguins. Love my little girls more than anything in the whole world. Love, Daddy."

Dr. Lewis told us that as soon as he repaired *Ice Bird* he wanted to sail around the Cape of Good Hope. Incredible, but true.

After two months alone on his disabled vessel Ice Bird, ***David Lewis*** *(left) sights* Calypso. *Over hot coffee (above) he tells of his plans to continue circumnavigating Antarctica when* Ice Bird *(below) has been repaired.*

To Prove a Point

In 1947, Thor Heyerdahl and *Kon-Tiki* gave support to the idea that man had crossed the Pacific from South America to Polynesia in rafts. In 1969 he set out to prove another theory—that ancient Egyptians had crossed the Atlantic to America on papyrus boats, using ocean currents as a guide, and that the similarity between the reed boats of the Egyptians and of the Mayans and Incas was no accident. Heyerdahl and six companions attempted to make this same voyage in a reed vessel named *Ra,* in honor of the ancient Egyptian sun god.

Ra was towed across Northern Africa from the shadow of the pyramids in Egypt, where it was built according to ancient specifications, to the port of Safi, Morocco. Here, Dr. Heyerdahl hoped an ocean current would carry the boat in a southwesterly direction through the narrow, but dangerous (due to offshore rock formations and a peculiar coastal outline) passageway between Africa and the Canary Islands. He believed that another current would then take over, carrying *Ra* across the Atlantic in a westward voyage. He thought the *Ra* might touch the Antilles after 45 days and arrive at the Yucatan Peninsula a few weeks later.

Many skeptics predicted that the craft would become waterlogged upon launching and sink, but it did not. But Heyerdahl and *Ra* never reached their destination, and it was for a reason nobody had given a second thought to—a lesson that taught us an extremely important facet of ancient Egyptian shipbuilding designs.

Ra's stern eventually became completely submerged, and Heyerdahl realized that what he and the crew had believed to be no more than a decorative detail was in fact an ingenious engineering device enabling papyrus boats to stay afloat on the open sea. All

such ancient vessels had a stern turned forward and upward, arching inward and fastened by a taut rope attached to the deck. (*Ra* had no such rope, because it was thought to be useless.) The rope served to keep the stern above water and to hold together under unusual stress the reed bundles of which the boat was constructed. By the time Heyerdahl and his crew realized this, the stern was so heavily saturated with water that it was impossible to ride over even the smallest waves. Eventually, Heyerdahl and his shipmates abandoned the *Ra* when it broke up on the high seas slightly more than three-quarters of the way through its transatlantic odyssey.

Undaunted by this setback, the crew became more determined than ever to reach their destination and to prove Heyerdahl's theory. Some ten months later *Ra II* was built by Indians from Lake Titicaca in South America who were brought to Morocco for the job. *Ra II* was 10 feet shorter than her predecessor and had a crew of eight instead of seven. More importantly, it had a stern that could not be vanquished by the sea. Closely adhering to the route taken by *Ra, Ra II* successfully completed the transoceanic voyage from Safi to Barbados in just 57 days.

Another man had a theory that one could survive for long periods with only the nourishment that the sea itself could provide. He was a French physician named Alain Bombard. In 1952 he set out in a rubber raft named *L'Hérétique* to prove his point. He planned to sail from Casablanca across the Atlantic to the West Indies. During the entire voyage he would eat only fish and use their juices and rainwater to provide the necessary liquids. Stowed in the raft were containers of emergency food and water rations. If the containers reached their destination with the seals unbroken, the expedition would be a success.

The first few days of the solitary voyage went well. "There were plenty of fish," he wrote. "Little flyingfish struck against my sail and fell in the raft." He varied his diet with plankton that he caught in a fine-mesh net. "It tasted like lobster, at times like shrimp, at times like some vegetable."

For 23 days he had no rainwater, but fish juices quenched his thirst. Still, he longed for great quantities of some liquid. "I dreamed of beer," he said.

Bombard discovered that the real difficulty was in what he called "the terror of the open sea." There was "no sound of fellow human beings or the familiar noises of the land. Only the rushing of the wind, the watery hiss of the breaking waves, the nervous flutter of the sail. And the face of the shark following you patiently, relentlessly—sometimes rubbing his sandpaper back under the raft."

Finally, after two months on the open Atlantic, *L'Hérétique* landed at Barbados. The emergency rations had not been opened.

In 1969, Thor Heyerdahl's first effort to prove that ancient Egyptians might have crossed the Atlantic Ocean in ***papyrus boats*** *ended in failure as* Ra *(below and left) was slowly overcome by the sea.*

Hero of the Antarctic

When Britain entered World War I, Ernest Shackleton, who had already made three voyages to Antarctica, offered to abandon his projected fourth expedition and place himself and his ship at the service of his country. The cable he got in reply said simply, "Proceed," and was signed by the First Lord of the Admiralty—Winston Churchill.

The expedition intended to make the first crossing of the continent—from the Weddell Sea to the Ross Sea—but it never got underway, because the ship was caught in the ice of the Weddell and crushed. Shackleton and his men found themselves stranded in the most remote of frozen seas with no way of telling the world of their plight.

By sled and sail, Shackleton and his men made their way to the nearest point of solid land—a hitherto untrod rock called Elephant Island. Leaving 22 men behind while he and five others went for help, they set out on one of the most daring voyages ever undertaken—a journey in a 20-foot open whaleboat across nearly 1000 miles of Antarctic Ocean in the remote hope of reaching a tiny inhabited speck in the South Atlantic called South Georgia Island.

The 16 days they spent in the *James Caird* were, in Shackleton's words, a tale of "supreme strife amid heaving waters." Once clear of the dangerous ice packs, they were hit by almost constant gales. Cramped in their narrow quarters and continually wet by the spray, they suffered severely from cold throughout the journey. There were no dry places in the boat, and bailing was a constant occupation. The incessant motion of the boat made rest impossible.

The meals were the bright beacons of their days. Breakfast consisted of "a pannikin of hot hoosh made from Bovril sledging ration, two biscuits, and some lumps of sugar." Lunch was comprised of the same sledging ration eaten raw and a pannikin of hot milk for each man. Tea consisted of the same menu. Then at night they had hot milk again.

When the ***ice of Antarctica*** *caught and crushed Ernest Shackleton's ship in the Weddell Sea, there began one of the most heroic struggles for survival in the history of polar exploration.*

A thousand times it seemed that their little boat would be engulfed by the seas, but each time she survived. Spray froze on the boat and gave everything a heavy coat of ice that constantly had to be picked and chipped at. The men were frostbitten and had large blisters on their fingers and hands.

At midnight of the eleventh day Shackleton was at the tiller when he noticed a line that appeared to be clear sky between the south and southwest. He called to the other men that the sky was clearing. A moment later he realized that what he had seen was not a break in the clouds but the white crest of the most gigantic wave he had ever seen in his 20 years at sea. When the wave struck, Shackleton wrote, "We felt our boat lifted and flung forward like a cork in breaking surf. We were in a seething chaos of tortured water . . . . Earnestly we hoped that never again would we encounter such a wave."

Their tongues were swollen by raging thirst, but their hopes were buoyed when they saw two shags sitting on a mound of kelp. They knew they must be within 10 or 15 miles of shore. "Those birds," Shackleton wrote, "are as sure an indication of land as a lighthouse is, for they never venture far to sea." The next morning they were hit by one of the worst hurricanes any of the men had ever experienced. "The wind simply shrieked as it tore the tops off the waves and converted the whole seascape into a haze of driving spray." The seas drove them toward reefs where great glaciers ran down to the sea. Then, miraculously, the wind changed, and they were able to find a landing in a sheltered bay.

But their journey was not over. They still had to make the first trek over what had been thought to be impassable glaciers in order to reach the whaling station on the other side of the island. But they succeeded, and soon after the men who had been left behind on Elephant Island were rescued.

"I'm the Captain. I'm Staying"

Fog shrouded the mine-sown Elbe River and the North Sea when Henrik Kurt Carlsen took his ship, the *Flying Enterprise,* out of Hamburg in late December 1951. In the Atlantic one of the most devastating storms of the century raged. The air waves were flooded with SOS calls from ships in trouble. Captain Carlsen's SOS would capture the attention of the world.

The ship moved slowly through the English Channel, barely holding headway against the rough water. Then it reached the Atlantic, where 30-foot waves assailed it. The ship was light, carrying only a third of her cargo capacity, and she would roll uncomfortably even in a moderate sea. On the day after Christmas, 200 miles from Land's End, Carlsen decided it would be safer to lay to until the storm abated rather than risk damaging his engines. The next morning he watched a mountainous wave crack his ship in half. The crew made a hasty patch, stuffing the crack with cement and binding the two halves together with wire cable turned around fore and aft bitts. Carlsen brought his disabled ship around and, working his motors at half-speed, attempted to get down to the shipping lanes where he would be more apt to find help.

On December 28 the sea dealt the *Flying Enterprise* a fatal blow. A 60-foot wave crashed on the ship, shattering wheelhouse windows and smashing a lifeboat. It left the ship listing 60° to port and powerless. Carlsen's SOS was answered by a dozen nearby ships. When the difficult and dangerous job of transferring his 10 passengers and crew had been accomplished, Carlsen went to the radio room "to square accounts with the ships around me." Asked when the captain would leave, Carlsen replied, "I'm the captain. I'm staying."

Alone on his ship, Carlsen took stock of his situation. A rescue tug was promised. He had faith in the integrity of his ship—the only leaks were through the open fire hose standpipes and it would take a long time for enough water to enter them to endanger the ship. Carlsen changed into dry clothing, selected a sleeping cabin on the low side, and made his bed in the V formed by the bulkhead and a leather-covered settee. There he had his first sleep in eight days.

As the *Enterprise* rolled on its lively arc, Carlsen attended to his paperwork, penciling his rough log on the cabin wall. He brushed up on the legal aspects of his plight by consulting a book dating from his cadet days in Denmark, *The Seaman and the Law.* When an airplane circled over his ship, Carlsen went out on deck and waved. One of the fliers took a photo of the ship and sent it that evening to a newspaper on Fleet Street. A sharp-eyed editor thought he saw a man on the abandoned ship and sent a reporter on a second flight to get the story. On New Year's Day the news was out and Carlsen was in another storm—a publicity hurricane.

During all of the 150 hours the ship had endured her crazy list, there had been ships standing by ready to take Carlsen off. Now Carlsen talked on his radio with the captain of the *John W. Weeks* and agreed that if his ship began to sink, he would set off blue flares and abandon ship on the low side and swim for the *Weeks.* Learning that Carlsen had been dining for six days on raisin pound cake, the *Weeks's* crew prepared a large bag of food for him. But Carlsen had no gloves and he was saving his hands for the difficult chore of receiving a tow cable. He refused to handle a cutting nylon line for food.

On January 3 Carlsen was awakened by sirens. He hauled himself up the slippery slanting passageway to the radio room. It was good news, the *Turmoil,* a powerful rescue tug, had arrived. "I propose passing a towline at once," her captain said. "Roger, Captain Parker," Carlsen replied.

The job of establishing a tow on a ship the size of the *Enterprise* usually requires six to twelve men. Carlsen got his hands on one of the many lines tossed to him that night, but the line parted before it could be secured. Just before noon the next day, Carlsen caught and held the heaving line and passed the tow cable back to *Turmoil.* The big wire was two feet from the tug's winch when the hawser parted and the wire screamed out through the fairlead of the *Enterprise.* They would have to try again.

The two vessels were so close that the tug's chief mate, Kenneth Dancy, was able to grab a dangling lifeline and join Carlsen on the *Enterprise.* On January 5 the two men finally succeeded in placing the ship in tow 240 miles from Falmouth Bay where they expected to arrive in four days. But on that day, January 9, the ship was 42 miles from haven, the sea was building up again, and the tow had been lost. On the thirteenth day of her ordeal the *Flying Enterprise* was turning over on her side and water began pooling in Carlsen's bedroom. Carlsen gathered some blankets and said, "Mr. Dancy, I think we'd better go out on deck."

In the raging storm it was impossible to send a helicopter for the men. On his last radio meeting with the tug captain, Carlsen said, "I think we had better come off the funnel." As a voice over the radio warned, "Don't wait too long," the men jumped and swam for *Turmoil.* There they watched the proud ship go down. The *Flying Enterprise* sank one minute before sunset on the fifteenth day of total disablement, after surviving two hurricanes and five storms.

For 13 days the captain of the crippled freighter **Flying Enterprise** *stayed alone with his ship as she wallowed helplessly in the sea off England.*

Chapter VII. Diving to the Past

Many of the answers to our questions about man's past lie hidden beneath the sea. They are in ships that have sunk and in the cargoes they carried. They are in harbors or entire cities that either sank beneath the sea or were inundated by rising sea levels. Most sites have been discovered by accident. A handful of them have been successfully excavated by archaeologists. A great number of them await investigation. An incalculable number of them wait to be discovered.

We have already learned a great deal about ship construction and trade routes, art (especially sculpture), technology, and history from the work of the archaeologist underwater. New knowledge of ancient metallurgy, numismatics, the science of weights and measures, and architecture, among many areas of human activity, has been gained.

> **"Looting has been practiced for centuries, but in recent years it has become rampant."**

Thus far, the Mediterranean has provided more sites than any other sea. Its floor is known to be littered with thousands of wrecks. This is understandable. The Mediterranean is probably the first large body of water on which man sailed. Freight of all description has been conveyed across it since about 3000 B.C.

The kind of sea a ship sinks in is crucial to its preservation. Fresh water does the least damage to wrecks. A ship raised from Lake Champlain between New York and Vermont after more than 150 years on the bottom was in excellent condition right down to the splinters around the hole made by the cannonball that sank her. Brackish water like that in the Baltic Sea does the next best job of preservation. A wreck from about 1700 lies there today with her lower masts still standing. The rotten ends of sheets and halyards are still hitched to the cleats inside the bulwarks. The most famous wreck of the Baltic is the Swedish warship *Vasa,* which was raised and placed in a specially built museum in Stockholm. The ship and nearly everything that had been in it was in a good state of preservation, and it had sunk more than 300 years ago. The Black Sea, with salinity at diving depths about half that of the Mediterranean, is sure to hold many well-preserved sites.

The future of marine archaeology is closely tied to advances in diving and underwater technology and also to education and to protection by governments of underwater sites. Looting has been practiced for centuries, but in recent years, with the enormous increase in the number of sports divers, it has become rampant. It is probably safe to say that there are no visible wrecks on the Spanish, French, or Italian coasts under less than 150 feet of water that have not been ransacked and almost totally destroyed. The Caribbean, too, has been a favorite haunt of plunderers for many years. Sporadic efforts to provide an education in the value of these underwater treasures and a few national antiquities laws have begun to provide a measure of protection, but far more needs to be done. The challenge of marine archaeology is not to the diving scientist alone but to the value every man puts on rational inquiry.

*The floor of the Mediterranean is littered with the remains of thousands of **shipwrecks.** Here, a diver reaches out to touch an amphora, often the only clue to a site of an ancient shipwreck.*

Amphoras and Cannons

From antiquity, man has sailed the sea. Archaeologists, working underwater, uncover remnants of centuries-old shipping disasters.

Calypso *divers (A & B) investigate an* ***ancient shipwreck*** *in the Mediterranean Sea.*

The ***Caribbean Sea*** *offers the remains of Spanish galleons (C, E, G, H) long ago sent to the bottom.*

Remains of a 2000-year-old ship *(D & F) were uncovered by marine archaeologists off Kryenia, Cypress. Much care was required to excavate it.*

▲C

▼A

▼D

▼B

▼E

Amphoras and Cannons

From antiquity, man has sailed the sea. Archaeologists, working underwater, uncover remnants of centuries-old shipping disasters.

Calypso *divers (A & B) investigate an **ancient shipwreck** in the Mediterranean Sea.*

*The **Caribbean Sea** offers the remains of Spanish galleons (C, E, G, H) long ago sent to the bottom.*

***Remains of a 2000-year-old ship** (D & F) were uncovered by marine archaeologists off Kryenia, Cypress. Much care was required to excavate it.*

▼A

▲C ▼D

▼B

▼E

▼F
▼G
▲H

The Sunken Museum

In the spring of 1900 a Greek sponge boat was riding out a storm in the shelter of the tiny Greek island of Antikythera when one of the divers decided to try to profit by the delay. He went down to look for sponges and instead discovered one of the most important archaeological sites in the Mediterranean.

What he found was a large field of partially buried bronze and marble figures, and he brought up a bronze arm as proof that he had not been dreaming. It was not long before an official salvage expedition was organized. The methods of the sponge divers who carried out the salvage were primitive. They groped about for objects which they tied with a rope and hauled to the surface. More than once the rope broke and the objects tumbled back onto the bottom.

Among the artifacts raised from the bottom were a bearded head, a pair of fifth-century-B.C. statuettes, and a bronze bed decorated with animal heads. Less well preserved were 36 statues, 33 fragments of arms and legs, and four horses, all in marble. Many of the items have been identified as copies of originals, and they are important to our knowledge of the beginning of exact copying techniques. The Romans were so fond of Greek art that the supply of original works had begun to diminish.

In addition to works of art found at Antikythera, there was a mechanism of gear wheels, dials, and plates. It was later identified as a complex astronomical computer by which positions of the sun, moon, and planets could be calculated.

It has been determined that the ship sank between 80 and 65 B.C. And it has been possible to determine a great deal about the ship itself. We know that it was copper-fastened and was built shell-first. She was probably decked and weighed about 300 tons. A tiny piece of wood from a plank was subjected to carbon-14 dating, and it was found that it was from a tree that had absorbed its carbon-14 between 260 and 180 B.C.

Marble sculptures *from Antikythera (left) are irreparably pockmarked by marine organisms.* ***Amphora*** *(above) dates from about the same age.*

The hull still lies there, 180 feet beneath the surface, beside a great weed-covered boulder beneath about three feet of sediment. No drawings were made of the site and no photographs were taken, so we can only guess at how it must have looked. We should not blame those who were there, for it was the first excavation of a Mediterranean site.

The Archaeologist Joins the Diver

At the time of Odysseus's epic voyage, the Phoenicians had already begun their famous trade throughout the Mediterranean Sea, but no one knew this until marine archaeology told us. The excavation of a Bronze Age shipwreck off the southwest coast of Turkey in 1960 provided the evidence that the Phoenicians traded by sea as early as 1200 B.C. It was the first underwater excavation carried out methodically and to completion, and it was the first time that an archaeologist played an important role as a diver. He was Dr. George F. Bass of the University of Pennsylvania Museum.

The site was near Cape Gelidonya off the southwest coast of Turkey. The ancient ship must have run against a small island, for it had settled between the base of the island

At base camp, *George Bass and Peter Throckmorton study a marine chart for navigational obstacles that may have doomed ancient ships.*

and a huge boulder, 90 to 95 feet beneath the surface. Sometimes there is more than sand and mud to be removed from an underwater site, and that was the case at Cape Gelidonya. Coralline concretion, hard as cement, covered most of the site. At places it was as much as eight inches thick, and its removal was a major problem in excavation. Almost nothing was visible except the ends and corners of a few metal objects.

A free diver made the first of a number of photographic montages of the site, and the positions of visible artifacts were plotted on sheets of frosted plastic. The entire cargo was raised in coralline masses and excavated on land. Divers with hammers and chisels cut massive lumps free from the seabed and then they were winched to a boat on the surface. Pieces of concretion containing parts of the wooden boat were raised to the surface with the aid of two plastic lifting bags. The wood was fragmentary, but it contained pieces of planks with tree nails fitted into bored holes. The interior of the hull was lined with brushwood dunnage, as described by Homer, and the bark was still well preserved.

There finally emerged a picture of a small sailing vessel, about 35 feet long, that carried more than a ton of metal cargo. The cargo consisted largely of copper ingots, many of them stamped with signs in a language that still has not been deciphered—Cypro-Minoan. Bronze implements found included hoes, picks, axes, and knives. Most of them had been broken in antiquity and been packed into wicker baskets. On board too were pieces of casting waste, as well as tools that had been cast but never hammered or sharpened. The cargo, then, was a load of scrap metal that had been destined to be forged into new weapons or tools.

Items were also found that evidently belonged to the ship itself or to its crew. They included a seal for stamping official documents, five scarabs, balance-pan weights, traces of food, including olive pits, stone maceheads, a razor, whetstones for sharpening tools, and an oil lamp.

Radiocarbon dating of the brushwood used in the hull revealed that it was from about 1200 B.C. Studies of the pottery gave the same dating. Examination of the bronze implements showed that most of their prototypes were found in Syria and Palestine earlier than in Greece. This indicated that either Canaanite or Phoenician merchants were responsible for the cargo and the ship.

Recovered objects *are placed on land exactly as they were found in wreck (above). Position of objects is recorded before removal (below).*

Among the artifacts discovered at Mahdia were a number of marble ***Ionic capitals.***

When this heavily encrusted cylinder was cleaned, it was found to be a fluted marble ***Ionic column.***

An Underwater Museum

One day in 1907 an itinerant Greek sponge diver reported to the commander of the French Tunisian naval district, Admiral Jean Baehme, that he had found a place where the sea floor was covered with cannons. Baehme sent his divers down to investigate. They brought back a heavily encrusted cylinder, cleaned it, and found a fluted marble Ionic column.

The discovery caused much excitement, and several wealthy patrons subscribed to a salvage effort. The Mahdia galley yielded enough objects to fill five rooms in the Museum Alaoui in Tunis. There were marble columns, capitals, and other architectural units, marble and bronze statues, bronze candelabra, and huge kraters. Scholars attributed the art to the first century B.C. The ship was probably Roman, transporting the spoils of Athens after its defeat in 86 B.C.

The wreck was at a crucial depth of 127 feet; some of the Italian and Greek hard-hat civilian divers suffered crippling bends. In 1913, after five years, Lieutenant Tavera, the leader of the operation, closed it out for lack of funds and deposited his report in the Tunisian archives.

Thirty-five years later, as a member of the Group for Undersea Research of the French Navy, I read Tavera's report. I was touched to find Admiral Baehme's name—he was a grandfather of my wife. We resolved to find the wreck again and to recover treasure we felt was left behind by the earlier salvors. We would have the advantage of the aqualung and new diving tables that cut our decompression stops drastically.

First, however, we would have to find the wreck. Time had obliterated all of Tavera's carefully defined landmarks. We had to start from scratch, knowing only the water depth. Our divers swam to and fro over a grid 100,000 feet square. Then we towed them over larger areas. For five days we found nothing. On the sixth, as we were becoming anxious about how we could justify a fruitless search for an already excavated wreck, Philippe Tailliez found a column.

We began a businesslike attack on the Roman ship. Teams of two went down and were rigidly timed. A rifleman on deck fired shots into the water that the divers could

The wreck at Mahdia yielded enough ***objects of art*** *to fill five rooms in the Museum Alaoui in Tunis.*

Art historians attribute the ***works discovered at Mahdia*** *to Greece in the first century B.C.*

hear; the first at five minutes, the second at ten, and at fifteen minutes three shots commanded them to the surface.

We didn't find the masterpieces of Greek art we'd dreamed about, but in six days of diving we recovered columns, capitals, and bases that made up the bulk of the large ship's cargo. We scrubbed the encasing mud from one Ionic capital to reveal marble as flawless as it was on the day it was carved. Of the ship itself, we brought up yard-long sections of Lebanon cedar ribs, iron and bronze nails, and two lead crossbars that gave horizontal stress to its wooden anchors. We estimated that the ship was still two-thirds intact under its cover of mud. We touched her solid, lead-plated deck briefly before the mud poultice spread over it again.

The Overdue Cargo

Calypso bobbed at anchor dangerously close to a desolate, craggy white rock a few miles from Marseilles. It was 1952, and the rock was named Grand Congloué. In the water I was searching for a large mound of amphoras I'd been told was there. Amphoras in a pile would be a sure sign of an ancient wreck.

At a depth of 220 feet, I found nothing. I swam to another side of the cliff and rose to 200 feet, where I spotted a lone amphora. But a single jar is not proof of a wreck. My air was running out, and I'd begun to feel I'd been misled. But when I'd risen to 140 feet, I saw it: a high pile of sand and pottery rubble. I worked loose three black cups.

Aboard the ship that day was a special guest, Professor Fernand Benoît, director of archaeology for Provence. When he saw the cups, he recognized them immediately. "They're Campanian!" he exclaimed. He'd seen shards of the same second-century-B.C. pottery recovered from land digs. They dated the Grand Congloué wreck as the oldest one that had been found to that time. Excavating the wreck would shed light on the life of men who lived 2100 years ago. We decided that we'd tackle it. We didn't realize that it would take five years and be the most extensive underwater archaealogical operation yet attempted.

The wreck was a big one, approximately 110 feet long and with a cargo capacity of 350 tons. It was in water depths of 125 to 145 feet. Our divers used a "vacuum cleaner" made of a large hose which was fed compressed air by a smaller one to create suction and lift away mud and small objects from the site. The hose emptied into a collection basket on Grand Congloué where small artifacts were caught. The salvors filled another basket with freed pottery and ship parts to be lifted and sent to the Borley Museum in Marseilles.

More than 7000 amphoras were recovered. The jars were of two types—squat Greek ones and slender Italian ones. They were used to carry wine, oil, or grains. As we unloaded them, we carefully noted their location in the wreck. It would help us to piece together the ship's last journey. Many of the Greek jars bore a seal with a trident and the letters SES. Benoît was able to establish with this clue that the ship had been

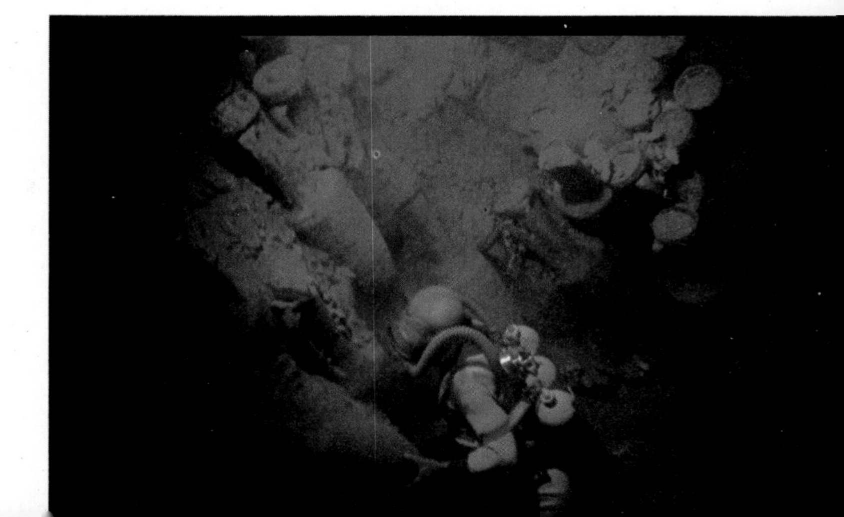

owned by a Roman named Marcus Sestius, who had lived on the Greek island of Delos at the time of the sinking.

To retrace the final voyage, we took *Calypso* to Delos. There we found in mosaic the letters SES worked into a design with a trident. The mosaic was to have been the floor of an unfinished villa. In our imaginations we can picture the powerful Marcus Sestius sending out his ship laden with wine in the fat amphoras of Delos. Stopping in Syracuse, it loaded more wine in Italian jars. At Naples it took on more than 7000 pieces of fine Campanian dinnerware. But perhaps a mistral was blowing as the ship approached Massalia, and it was tossed against the rock. And with the loss of his ship and his fortune, perhaps Marcus Sestius could not afford to complete his villa.

The ***amphoras*** *discovered at the site of the ancient shipwreck off Grand Congloué (below and opposite top and bottom) helped reconstruct the ship's last journey. The inscription on some of the jars (opposite center) was the same as that found at the partially constructed villa of a wealthy Roman who lived on the Greek island of Delos.*

Chapter VIII. The Sea That Divides Men

In ancient times naval battles were fought by ramming the enemy ship and then boarding her and fighting hand to hand. The greatest battle of antiquity was Salamis, where about 1000 Persian ships were defeated by some 300 Greek galleys, or triremes.

When the Roman Empire collapsed, the Norsemen became masters of the seas, at least the northern seas. The earliest northern sea battle was between the Vikings and the English. King Alfred of England had built larger boats than the Viking longboats, and his fleet of dragon boats met the enemy at sea instead of waiting for them to land. During the entire second half of the sixteenth century a state of war at sea existed between England and Spain. It all came to a climax in 1588 with the defeat of the Spanish Armada. It was the first battle at sea to be fought with guns alone. In the nineteenth century, Nelson's victories over the French changed the course of naval history. Until his time, naval forces engaged each other in parallel lines close enough to fire broadsides. But at Trafalgar he came at right angles and split Napoleon's fleet.

"By the end of World War II the submarine had come to dominate the war at sea."

During the long years of relative peace between the Napoleonic Wars and World War I, the navies of the world were in a turmoil. Wooden sailing ships had been rendered obsolete by the advent of steam, the introduction of iron and steel, rifled guns, and the explosive shell. By the early 1900s a naval arms race had begun. The British made a major advance when they built the *Dreadnought* in 1906. The ship was able to carry more big, long-range guns by eliminating its secondary armament. Germany speeded up her program and soon had ships of the same type. For the four years of World War I massive fleets of dreadnoughts threatened each other, but only one major battle took place—the Battle of Jutland—and the outcome was indecisive.

By 1900 the principle of the submarine as a weapon of naval attack had been generally accepted. Now, in World War I it became a major weapon. In two and a half months and with only 20 submarines, the Germans sank nearly 100 ships.

In World War II the Battle of the Atlantic was in many ways a continuation of the U-boat battle of 1914–18. In 1942 America and Great Britain were losing about 100 ships a month to submarines.

In the Pacific it was an entirely new kind of war at sea. It started when 350 Japanese aircraft took off from six carriers to attack the American fleet at Pearl Harbor. The attack, which almost wiped out the U.S. Pacific Command, set the pattern for the war in that ocean. The naval war was to be fought between unseen fleets launching massive air attacks on one another. Carrier battles raged across the Pacific, and for a time aircraft were the greatest power at sea. But all this time, much less dramatically, American submarines were sinking nearly the entire Japanese merchant fleet and a few of her greatest warships. By the time that war was over, everyone knew that the submarine, not naval aircraft, would dominate the seas in the future.

A propeller, *encrusted with marine creatures, is one of the few discernible objects amid the twisted wreckage of a vessel sunk in an unknown tragedy.*

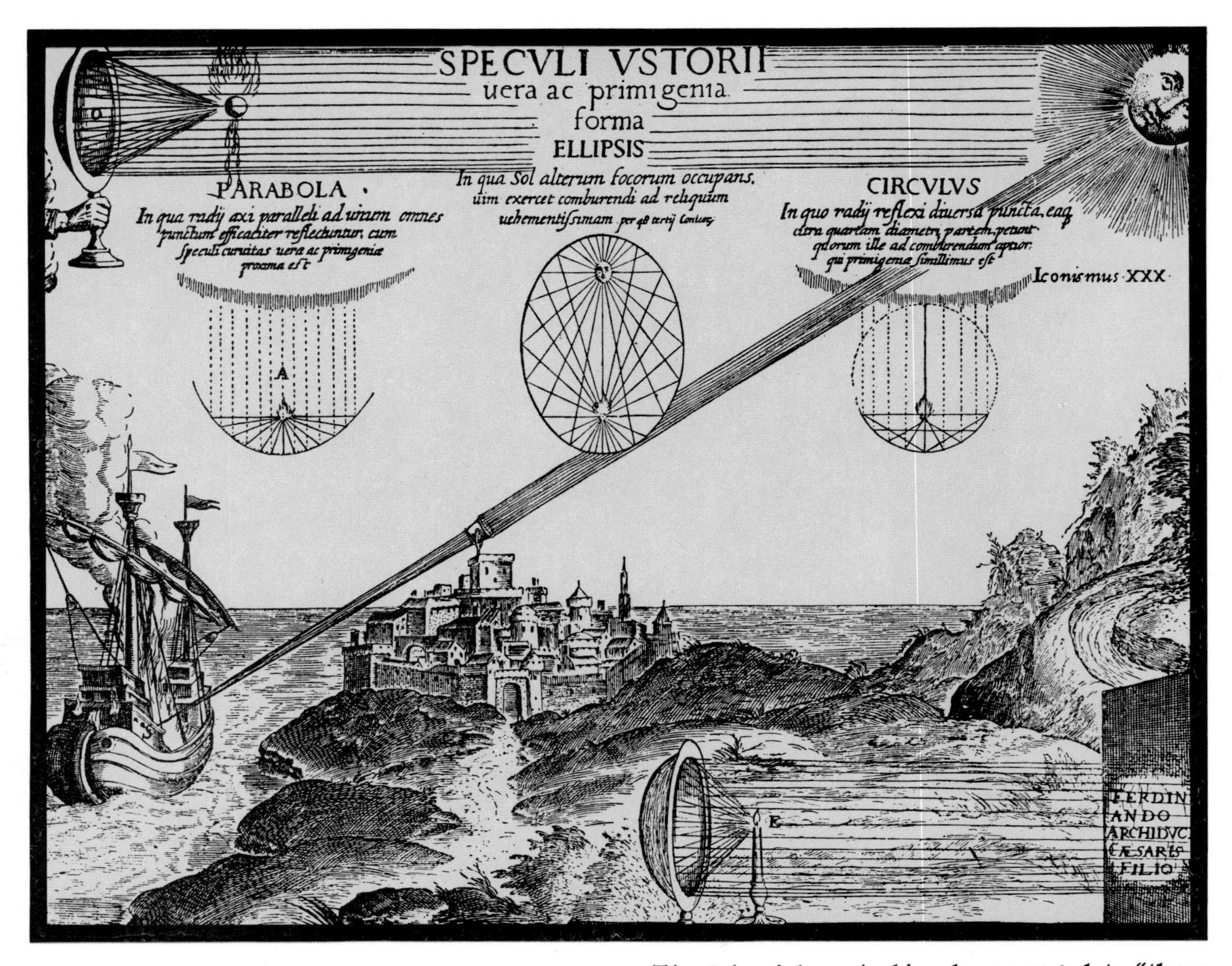

The tale of how Archimedes managed to "throw fire" in order to destroy the Roman fleet attacking Syracuse has long puzzled classical scholars. Artists had no problems in depicting ***fanciful versions*** *of how the phenomenon could have occurred.*

How Fire Was Thrown Across the Water

Enemies that are separated by water traditionally feel secure in the knowledge that they are safe as long as there are no boats on the surface or any underwater activities. But the separation that the sea provides is sometimes illusory. Water, for example, has always been assumed to be a protective barrier not only for countries, but also for ships that are blockading a harbor or massing for an attack. By maintaining a distance, the saying goes, safety is maintained.

But distance is a relative thing, for what is out of range for one weapon, such as an arrow, may well be within the capability of another. It took not a soldier, but antiquity's master philosopher, mathematician, and inventor, Archimedes, to demonstrate this. Legend has it that he used some sort of "burning glass" to set fire to Roman ships that were attacking his native city of Syracuse on Sicily between 215 and 212 B.C. How Archimedes had achieved this, however, remained a puzzle until 1973, or for about 2200 years, when an engineer named Ioannis Sakkas reproduced what he believes was the most plausible method.

The key is several large mirrors which all beam reflected solar radiation in the same direction. Archimedes, it is theorized, used about a 100 bronze mirrors trained on the Roman ships and set them afire although they were probably 130 to 160 feet offshore. "At first," Dr. Sakkas said, "I thought Archimedes had invented a large convex lens that he used to focus the sun's rays on the bows of the wooden galleys. I discarded it early because it would have been an impossible feat. We built 100 flat, bronze-coated mirrors measuring 5 feet by 3 feet."

With Greek sailors holding the mirrors along a pier in Athens, the rays were focused on a small rowboat 165 feet away which was made of slow-burning tarred plywood. Yet the concentrated solar radiation produced smoke from the boat within seconds and flames appeared within two minutes.

The Romans found out the hard way that the sea which separates may not be effective if the sun is shining.

In an effort to ascertain the accuracy of ancient stories, Dr. Ioannis Sakkas used ***100 men holding bronzed mirrors*** *to reconstruct the method by which Archimedes might have used reflected solar radiation to set Roman warships on fire.*

*One of the most memorable battles of the American Civil War and one of the most significant in naval history was between **the* Merrimac *and* Monitor.**

A Clash of Ironclads

On a Sunday in March 1862, a year after the American Civil War had begun, two ships clashed at Hampton Roads, Virginia, in a battle that made every navy afloat obsolete. Steam and armor plate had been combined in a new kind of warship, the ironclad.

The *Merrimac* had been scuttled by the Union when its forces had to abandon the navy yard at Norfolk, but she was raised by the Confederates and overhauled. Her masts and almost all the superstructure were removed. She was given a sloping roof covered with iron thick enough to deflect any cannonball that could be fired at her. The funnel rose out of the roof like a chimney, and ten square openings were cut in the armor as ports for the ten big guns—four on each side, one in the bow, and one in the stern. A sharp iron beak was fastened to the bow just beneath the waterline. It would prove to be a very effective ram.

Meanwhile, a Swedish engineer in the service of the Union was also working to combine the virtues of steam and iron in a new kind of fighting ship. The *Monitor*'s deck was also heavily armored, and her pointed bow was strongly reinforced to serve as a ram. A funnel rose from the low deck aft, and forward was a pilothouse made of iron beams. Amidships was a round gun turret with eight- and nine-inch iron plates. The gun turret represented yet another great stride in naval design and thus in naval tactics. The turret was turned by men working below deck so that the two 11-inch guns could be aimed in any direction without having to bring the ship around.

The Union leaders knew about the *Merrimac,* and they were badly frightened. The Confederate ironclad might be able to break a blockade by wooden ships. She might even steam up the Potomac and attack Washington or break into New York harbor and shell the city. When the *Monitor* was ready, therefore, it was dispatched with orders to meet the *Merrimac* and destroy her.

On the day the *Monitor* arrived at Hampton Roads, the *Merrimac* had already destroyed one Union ship and put several others out of action. She had plowed right into the *Cumberland* when that ship turned to fire a broadside and left a huge hole in her side that sent her to the bottom in less than an hour. Then the *Merrimac* turned to attack the *Congress,* and her captain, having witnessed the fate of the *Cumberland,* decided that the only way to avoid meeting the same end was to run his ship aground. Three other Union frigates—the *Minnesota,* the *Roanoke,* and the *St. Lawrence*—tried to sail away and got themselves stranded on a shoal. Then the *Merrimac* and the *Congress,* now high and dry, began to exchange fire. Cannonballs glanced harmlessly off the *Merrimac*'s armor, but the shells that struck the frigate soon had her on fire. At 7 P.M. the *Merrimac* withdrew. She would finish off the others the next day.

But she never got the chance, because in the night the *Monitor* arrived and the next day it was ironclad against ironclad. For hours their guns fired away at each other, but neither did the other any damage. At noon the *Merrimac* withdrew. No one had been killed or even seriously injured, but the battle had been one of the most decisive ever fought, for it was clear when it was over that the age of wooden warships was over.

Two months later the Confederates, having to abandon Norfolk, scuttled the *Merrimac.* And on the last day of the same year the *Monitor* foundered in a gale off Cape Hatteras. Their day was over, but the day of the ironclad had just begun.

The Merrimac *and the* Monitor *clash marked* ***the end of the day of the wooden fighting ship.***

Sowing the Sea with Death

A sharp-eyed lookout on the gunboat U.S.S. *Pawnee,* patrolling the waters of the Potomac in the early days of the American Civil War, saw two black specks bobbing in the water some 200 yards away. The vessel was stopped and a cutter launched to make an investigation. They found a pair of large barrels connected by a long rope. Suspended beneath the barrels were two iron containers filled with powder and fitted with waterproof fuses. The rumor that the Confederates had been experimenting with "explosive machines " was true, and here was the evidence. Water was poured into the airholes to extinguish the fuses, and then the mines were carefully towed toward shore, one of them sinking on the way. It was the first appearance of the weapon in the Civil War.

The mine had been experimented with in several instances of naval hostilities, but with no great success. During the American Revolution a "keg torpedo" had been devised by David Bushnell, but it was not successful. Floating mines were used at Canton, China, in 1857–58, and the Russians used them during the Crimean War. But not until the American Civil War did they come into use as a significant factor in naval warfare. In that war the mine destroyed or damaged 40 Union ships and a number of Confederate vessels and killed and maimed hundreds of men. Mines sent more Union vessels to the bottom than did all the warships of the Confederate Navy. Thereafter, they were used in every war in which naval operations played a part.

In the years immediately preceding World War II, the Japanese, like all the other great naval powers, still considered the battleship to be the backbone of sea power. They had built two monstrous ships, the *Yamato* and the *Musashi,* armed with nine 18.1-inch guns. Their mission was to do battle with other ships. Both of them were destroyed, not by other battleships but by naval aircraft. Appreciating then that it was not the battleship but naval air power that was to be supreme in the struggle for the seas, the Japanese hastily converted the still unfinished hull of a sister ship, the *Shinano,* from a battleship to an aircraft carrier.

On November 28, 1944, the *Archerfish,* an American submarine, was on patrol south of Tokyo. Its primary mission was to act as a "lifeguard" for any American plane that might be shot down on its way to or from a bombing mission over the Japanese main-

land, but when the *Shinano* hove into view, its mission became to destroy the enemy. Four destroyer escorts barred every approach. Unless the *Archerfish* could reach a position ahead of the *Shinano,* it would be suicidal to attack. Throughout the night and into the next day the submarine coaxed its engines to follow the prey. Suddenly, at three o'clock in the morning, the *Shinano* altered its course to SW, leaving the *Archerfish* dead ahead at six miles. The submarine slid beneath the surface.

On the sonar the pulsating rhythm of the carrier's propellers grew steadily louder. Finally, at 7000 yards, it came into view through the submarine's periscope. Conveniently, the nearest escort moved out of station to take a message flashed by the carrier. In doing so, she unmasked the giant ship at precisely the right moment. And a moment later the carrier made an alteration of course that left her a sitting duck for the sub. It was 3:17 A.M. Six torpedoes fired away at short intervals, and the submarine plunged to the depths. The first of the torpedoes ripped into the carrier's stern with a blinding explosion. Five other blasts followed one upon another. Each torpedo had found its mark.

The shock of the attack stunned the Japanese. The escorts loosed 14 depth charges, but none of them found the submarine. Their shock was hardly felt aboard *Archerfish* compared to the jolt of the great ship breaking up on her way to the bottom.

The *Archerfish* never sank another ship, and perhaps that was appropriate, for she stood at the end of one era and the beginning of another. Naval air supremacy gave way to the age of the submarine.

One of the most lethal of weapons is the ***missile fired from a submarine*** *(opposite top).*

A row of ***depth charges*** *stands ready for launching (opposite below) on a maneuver by the U.S. Navy.*

Depth charges *(below) continue to be one of the principal weapons in antisubmarine warfare.*

The Tombs of Truk

A number of Japanese ships and planes were sunk in Truk Lagoon during World War II. They still lie there, like a museum of the war in the Pacific that few have seen.

*A diver explores the **wreckage of Japanese ships** sunk in Truk Lagoon (opposite top).*

*Today, some relics of World War II, like the **gas mask** (lower left), are covered with silt.*

*The wreckage, with well-defined spaces (right), provides an **inviting habitat** for the lagoon fish.*

*Other relics like the **ship's wheel** (below, right) are overgrown with brightly colored coral and algae.*

▲B ▼C

A Garden of Shipwrecks

The sea is the burial ground for ships of trade, pleasure, and war. The sea has put all of them to a new use—providing an environment in which its own life can flourish.

***Unnamed wrecks** in the Caribbean (A, D, E) and South Pacific (B, C) lure divers and marine life.*

▼A

▲D
▼E

Chapter IX. The Sea That Unites Men

For tens of thousands of years boats were no more than fallen logs paddled by hand. Then one day someone thought of tying two of them together, and so he built the first raft. In another part of the world a man discovered that an armful of reeds was enough to keep him afloat.

Three basic types of boats soon began to appear all over the world. One was the dugout canoe—a log hollowed by fire or ax. Another was the raft, made from wood or reeds. A third was the skin-covered boat that probably developed from a bundle of reeds covered with an animal skin.

These primitive boats gradually changed. The rafts became saucer-shaped, which helped keep their crews dry. Then they became longer, which made them faster and easier to steer. Wooden paddles were found to be more efficient than hands. A long wooden paddle was used at the stern and thus the rudder was invented.

"For tens of thousands of years boats were no more than fallen logs paddled by hand."

The Egyptians, who had invented the sail and the keel and who for centuries dominated the eastern Mediterranean in their ships made of long cedar planks, lost their lead to the Phoenicians, who developed a shorter, broader ship with one large square sail and with it became the great traders of ancient times. They opened up the entire Mediterranean to their trade and even sailed beyond it. The next advance was probably the most important of all—the oarsman, usually a slave, was replaced by developments in the use of sail and in ship design. Then for hundreds of years the wind did all the work.

In the early years of the nineteenth century the steamship was invented. It did not immediately replace sail. In fact, steam was at first used only for entering and leaving harbors and in times of calm. Even when the much more efficient propeller replaced the paddle wheel on the steamship, sail advocates remained staunch, particularly when the fast clipper ships arrived on the scene. But as oil and coal became more available, even the clipper ships began to lose favor. Steamships, using inexpensive fuels, became more dependable than ships relying on wind.

Coaling stations were established all over the world, and the tramp steamer came to dominate the sea lanes. It could carry any kind of cargo. With the great increase in world trade, specialized cargo ships were built. The most important of them was the tanker. There had been romance in the tramp steamer, but there was none in a ship that carried oil.

The glamor of ships passed to the ocean liner in the twentieth century. Great ships like the *Mauritania*, the *Normandie*, and the *Queen Mary* were luxurious floating hotels, and a transatlantic voyage was a delightful experience. Then the airplane and our impatience changed all that. The last ship designed exclusively as an ocean liner for the transatlantic passenger trade was the *United States*, built in 1952. The airplane carries passengers, but not freight, more efficiently than a ship, and since World War II the oceans have become more important than ever before as avenues of trade.

*International trade serviced by great fleets of **oceangoing cargo vessels** joins busy port cities of the world's maritime nations.*

SUZUE WAREHOUSE

Navigating Salesmen

The Phoenicians were the first people to fully realize the potential of the sea as an avenue of trade. Their province was the Mediterranean. The Minoans had been successful traders on the Inland Sea before them, but when their power was destroyed around 1200 B.C., the Phoenicians opened it up from one end to the other.

Living on the eastern rim of the Mediterranean in an area 225 miles in length but never more than 12 miles in breadth, they found it increasingly difficult to feed a growing population. They would have to find food elsewhere, and give something in exchange for it. That commodity was in obvious abundance. It was cedar, and it grew in enormous forests in Lebanon. A small trade in cedar had already been established with Egypt. Now the Phoenicians took up that commerce and developed it.

Although the Phoenician seaboard was not well suited to navigation, being frequently buffeted by onshore winds, it did possess two good harbors at Tyre and Sidon. These became the principal ports of the mercantile nation. They began to trade with every people that wanted to do business with them. The trinkets that they made came to be in

great demand by Greek women, and the products of their looms were bought by the Greeks for their most important sacrifices to the gods. And of course, they traded their famous purple dye obtained from the murex shell. They followed an old Minoan sea-lane from Sicily to Spain and then they searched out the route along the African coast. From Spain they carried great quantities of silver. Wherever they discovered good harbor sites they created colonies—Cadiz in Spain, Valletta in Malta, Bizerte in Tunisia, Cagliari in Sardinia, and Palermo in Sicily. Carthage, near the present site of Tunis, was the most vital of its colonies.

An ***Arab trading ship*** *is loaded (above) much as it was hundreds of years ago when the* ***merchant vessels of the Phoenicians*** *(left) plied the length and breadth of the Mediterranean Sea.*

As explorers, they were second to none. In their ships with one square sail they sailed through the Red Sea and around Africa, and they voyaged from the Mediterranean through the Strait of Gibraltar to the west coast of Africa, as well as to France, England, and Ireland.

There are many mentions of the Phoenicians as traders in the Bible, and Homer refers to them on several occasions. In the *Iliad,* at the funeral games of Patroclus, one of the prizes is a silver bowl brought oversea by the Phoenicians.

The geographical environment led the Phoenicians to develop as they did. The configuration of their land had made it necessary for them to seek a sea outlet. They had met the challenge successfully. A Spaniard writing of them in Latin in the first century described them in a way corroborated by recent studies: "The Phoenicians were a clever race, who prospered in war and peace. They excelled in writing and literature, and in other arts, in seamanship, in naval warfare, and in ruling an empire."

The clipper ship was the fastest sailing vessel ever built and, during the 1800s, helped bring continents closer together. These majestic ships were eventually replaced by steam-driven vessels.

Trade Wings

When the steamship made its appearance in the middle of the nineteenth century, there were some who were not convinced that the days of the sailing ship were over. Their argument was bolstered by the appearance on the scene of the clipper ship, for here was the fastest sailing vessel ever built—one to rival those early steamships.

The clipper ship evolved gradually as an answer to the appetite of the time for speed. One of the first of the clippers was designed by an American marine architect, John W. Griffiths, who had persuaded some wealthy businessmen that such a ship would make money in the China trade. It was called the *Rainbow*. Her maiden voyage in 1845 was not a success—the crew was not used to the demands of a sail-loaded clipper.

Clippers with their considerable speed were useful in the mid-nineteenth century in carrying East Coast passengers to the Gold Rush in California. They made Atlantic crossings in less than two weeks at what seemed phenomenal speeds. The "greyhounds of the sea" soon dominated the transatlantic passenger trade and the immigrant route from Britain to Australia.

The *Cutty Sark* was one of the most famous of the clipper ships. Built in Scotland in 1868, she first carried tea from China, often getting home 10 days before her rivals. Later the *Cutty Sark* entered the Australian wool trade and once came home in only 69 days, when the average voyage took 100 days.

There were disadvantages to the clippers, however. The sleek streamlined ships had very limited cargo capacity. To make the speed for which they were famous, an enormous amount of canvas had to be spread, day and night, fair weather or foul. That took a large, rough crew, and a captain had to be a superman to command them. Then cheap coal became available and steam engines were built that were more practical. The opening of the Suez Canal in 1869 shortened the route from Europe to the Far East, and for European sailors the day of the clipper was over. They lasted longer in the United States, where they were modified to hold more cargo and require smaller crews. But with the coming of the twentieth century, steam won out on the high seas.

Obese Titans

Every day in 1972 the world consumed about 1.5 billion gallons of oil. Most of that oil must be transported in ships, and the only ships that are capable of satisfying the enormous need are the modern supertankers.

A supertanker is defined as any ship used for the transport of oil that weighs over 100,000 dead-weight tons. There are hundreds of such ships in operation today. Experts predict that before long they will weigh as much as 1 million tons.

Supertankers require superports. There are about 50 such ports around the world, and new ones are being constructed. With superports, fewer tankers will go to fewer places, and the chance of oil spills will be minimized. When a supertanker does have an accident, the damage can be devastating. The *Torrey Canyon,* an 118,000-ton tanker, sank off the coast of England in March 1967, spilling some 30 million gallons of oil that washed onto English and French beaches. Damage was estimated at more than $16.6 million, not counting the incalculable damage done to life in the sea and at the shore.

In order to minimize accidents, officers of supertankers are studying the action of wind, waves, and currents on models of the ships on a miniature ocean created for the purpose high in the French Alps. It is, in fact, only a nine-acre lake, but a standard seashore jetty has been built at one end, and deep-sea mooring buoys have been set up. Sea anchorages have been marked, navigation buoys have been placed, and in the middle of it all is a giant wave-making machine—all on a scale of one to twenty-five.

The Jules Verne, *a* ***supertanker*** *designed to carry liquified natural gas, plows through a calm sea. Such vessels require superports; most harbors are not deep enough to accommodate them.*

Chapter X. Challenges of the Future

One of the greatest challenges of the future might be, very simply, to find out what the sea is and what its relationship with mankind should be. We can do little to reap the harvest of the sea or minimize its destructive forces until we understand it far better.

Understanding begins with feeling. It is important that we know what the sea looks like, what it sounds like, what it smells like. We cannot bring everyone to the sea, but we can bring the sea to them—not literally, but in books, films, television, and various forms of art. The sea must first become a significant part of our culture. Only then will we be able to make the best use of it.

Some people have always lived in an intimate relation with the sea. Most of them are fishermen or sailors. Some are among the millions that live at the edge of the sea. A few of them are those who have always lived on the sea itself in houseboats. A handful of scientists has come to know the sea well, but a very few have explored it beneath its surface. For most men the sea is alien. They are not consciously affected by it. They give it no emotional thought. It is essential that they do. We are threatened on many fronts and the sea can save us. The sea can provide some of the food to feed our multiplying populations. It can provide the energy source to power our homes and industries. It can provide some of the minerals that we will soon deplete in their terrestrial deposits. It can inspire peace.

"The sea must become a significant part of our culture. Only then will we be able to make the best use of it."

Many people speculate on man's living on or in the sea. It may come about but not in the near future. To live on the sea is very expensive and not very practical. To live in the sea would be denying the very real fact that we are terrestrial creatures.

If we do come to live on or in the sea, the reason may not be overpopulation but rising seas. We are living right now in an age when the sea is claiming the land as it has several times in the past. We may be able to reverse the process by inducing glaciation and thereby locking up some of the waters of the oceans in ice. If this is not possible and the seas reclaim large areas of the earth's land, we will probably find that the solution is to live upon the sea.

We have already found that we are physically able to live under the sea for long periods of time, and we are now beginning to make use of underwater habitats for science, education, and recreation. A child may spend one summer's vacation finding out what a forest feels like and the next summer finding out what the sea feels like. But he will only be one of a very fortunate few. Day-to-day living under the sea will be limited to those who work on undersea mining operations, drilling rigs, and aquatic farms.

The sea has a precious educative value—standing alone on the shoreline, filled with wonder, a teenager may hold a pebble in his hand and think—the sea gave this stone its shape. How must it shape my whole life?

A diver uses a scooter *to give him added maneuverability as he explores the ocean bottom. One of the greatest challenges of the future will be to learn much more than we now know about the sea and what our relationship with it should be.*

Antiques of the Future

The square-rigged ships that survived to sail in the twentieth century usually set out with few hands and cheap cargo. The ships were slow, and the crews had to handle everything, for there were no engines aboard. The windjammers worked in the grain trade between Australia and England until the century's third decade, when a worldwide depression put them out of business. The wind still blew free, but shipping ran on fossil fuels. Now we are running low on oil, and wise men are looking once again to the force of the wind.

In 1956 a civil engineer named Wilhelm Prölss, who had spent most of his professional career on aircraft design, put his mind to work on sailing ships. He turned out a design for a vessel he called the Dyna-Ship, and took it to the Schiffbau Research Center at the University of Hamburg for testing. The innovative design called for four or more masts 200 feet high. Sails would roll out from the masts on horizontal tracks on the curved stainless steel yards. Yards could be fixed, sails set, and masts rotated by a single man pressing a button. Planned for service on the North Atlantic, the vessel

The sailing ship of the future. *Totally automatic and computerized, this vessel has a small crew and uses satellites for navigation and has masts and sails that rotate to make full use of wind.*

would be capable of 12 to 16 knots average speed, compared to the 10 to 15 knots averaged by diesel-powered ships, and in strong wind conditions the Dyna-Ship might hit a top speed of more than 20 knots. In a calm sea, an auxiliary diesel could be used.

During six years of testing and refining the design of the Dyna-Ship, it was proved that with modern materials and the navigational and weather forecasting aids available, a wind-powered ship would be as able to maintain schedule as the fuel-powered vessels. The savings realized in manpower and fuel would be considerable. Development of the Dyna-Ship awaits financing.

Another imaginative solution to shipping is the submarine-blimp train. Liquid or semiliquid cargo, like crude oil or grain, could be pumped into large, strong neoprene sacks that can be connected in long trains. A nuclear-powered submarine would pick up a train of such blimps and tow them underwater like so many inflated balloons. The system would eliminate the dangerous transfer of oil to and from the ship. Tankers take on seawater as ballast when they've discharged their oil. When they flush the water before taking on a new load of oil, they empty it into the sea. It carries with it, as well, a significant amount of oil, the residue of the earlier load. The oil blimps would eliminate this problem. They could be folded up and sent back to the oil fields in small packages. The need for deepwater ports and constant dredging of rivers to allow large tankers to get to oil refineries would be alleviated. The oil blimps could be towed in a minimum of depth.

*A nuclear-powered submarine delivers a **train of undersea blimps** to a futuristic city beneath the sea. Such a unique system could have numerous applications in transportation and trade.*

Weather-Making

Some progress has already been made in controlling the weather, and much more is expected in the near future. We must realize, however, that weather does not consist of isolated phenomena, but is made up of many functional interrelated units that work together like a series of gears or pulleys. As our technology and understanding of nature increase, it is logical for us to try to minimize the destruction and discomfort that extremes of weather impose upon us.

The key to climate manipulation is probably the control of glaciation. By learning how to induce freezing and melting at the polar caps, we will learn how to influence climate everywhere. One phenomenon of weather that man would most like to be able to control is the hurricane. One possible way of attacking the great storm might be through its boundary at the surface of the sea. Most of the energy that maintains the hurricane comes from the latent heat transported into the storm clouds, following evaporation of warm ocean water. If that flow of latent heat

from the sea could be reduced, the intensity of the hurricane should also be reduced. This is what happens when the storm passes over the cooler waters of the North Atlantic and is dissipated.

Each year lightning does extensive damage to trees and buildings. One possible way of controlling it seems to be in causing thunderstorms to discharge slowly and thus to eliminate the very rapid strokes of lightning. Putting very fine metal strips into the clouds has been proposed as a way of doing this. The strips would act to draw off the electrical energy stored in the clouds a little at a time; of course, these strips might have unsuspected side effects and could be considered as pollution in some instances.

Hail is another destructive force that we would like to control. A hailstorm can wipe out a field of wheat in minutes. Most experiments have aimed at increasing the number of hailstones and so decreasing their size. This has been done with silver iodide seeding. Sometimes, however, seeding has worked the other way. We must learn more about the properties of various atmospheres if any kind of seeding is to be successful.

Hurricanes, lightning, and hail are a few of the more dramatic phenomena of weather, but not the most important. Far more crucial is the amount of precipitation. The losses brought about by a lack of rain have always been great. So have the losses from too much rain. We know now that cloud systems can be modified, but we need to know a great deal more about the atmosphere and its relation to the biological community before control of precipitation can be conducted with confidence.

Efforts to control the weather are aided by photographs like this one of ***the eye of a storm*** *(left).*

The effects of fog on transportation are being reduced by ***weather control techniques*** *(right).*

The Human Pipeline

It is possible that the man of the future will be able to take a train from New York to Paris. He will sit down in a comfortable lounge chair facing east, and in less than an hour he will arrive in the City of Light.

The human pipeline is not a pipedream, but a sound engineering possibility. The train will move silently through a huge 4000-mile-long vacuum tube laid on the bottom of the sea as a transatlantic cable. It will be powered by a linear electric motor with no gears. Metal guideways will "steer" the train, but the train will never touch them. Instead, it will speed along about one foot above the rails, being supported, guided, and propelled by powerful electromagnetic forces. Such systems have successfully been tested in a number of countries throughout the world.

During the first half of such a trip from New York to Paris, passengers will face forward as the train accelerates, and the only sensation of speed they will have is of being pressed back into their chairs. In a vacuum tube there is no practical limit on speed. Halfway to the destination, the train will begin to decelerate, and the passengers will be automatically turned to face the rear of the train. They will once again feel as if they are being pressed into their chairs.

At an acceleration rate one-tenth as great as the earth's gravity, the train would be traveling at 120 knots after one minute. After 30 minutes the speed of the train would reach 3600 knots.

The human pipeline would be extremely costly to build, but after the intial cost the running expenses would be low and the convenience and nonpolluting factors would be considerable. Building such pipelines could only be efficient if operated between widely spaced cities.

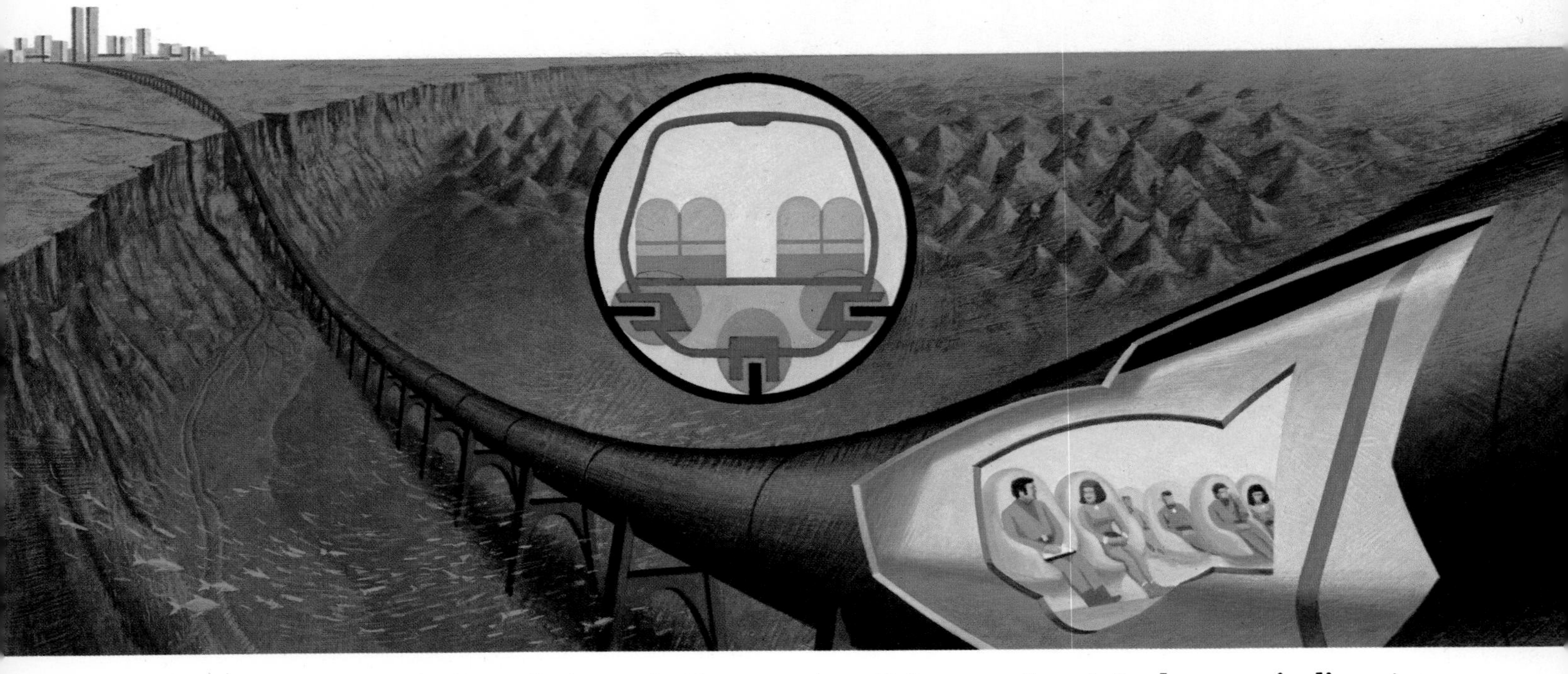

An artist's conception of the ***human pipeline.*** *A giant vacuum tube snakes its way down the continental shelf and slope to the ocean floor. In a vacuum, there is no practical limit to the speed the train could attain. In the inset above, the orange circles indicate the points of magnetic contact which would propel and suspend the train.*

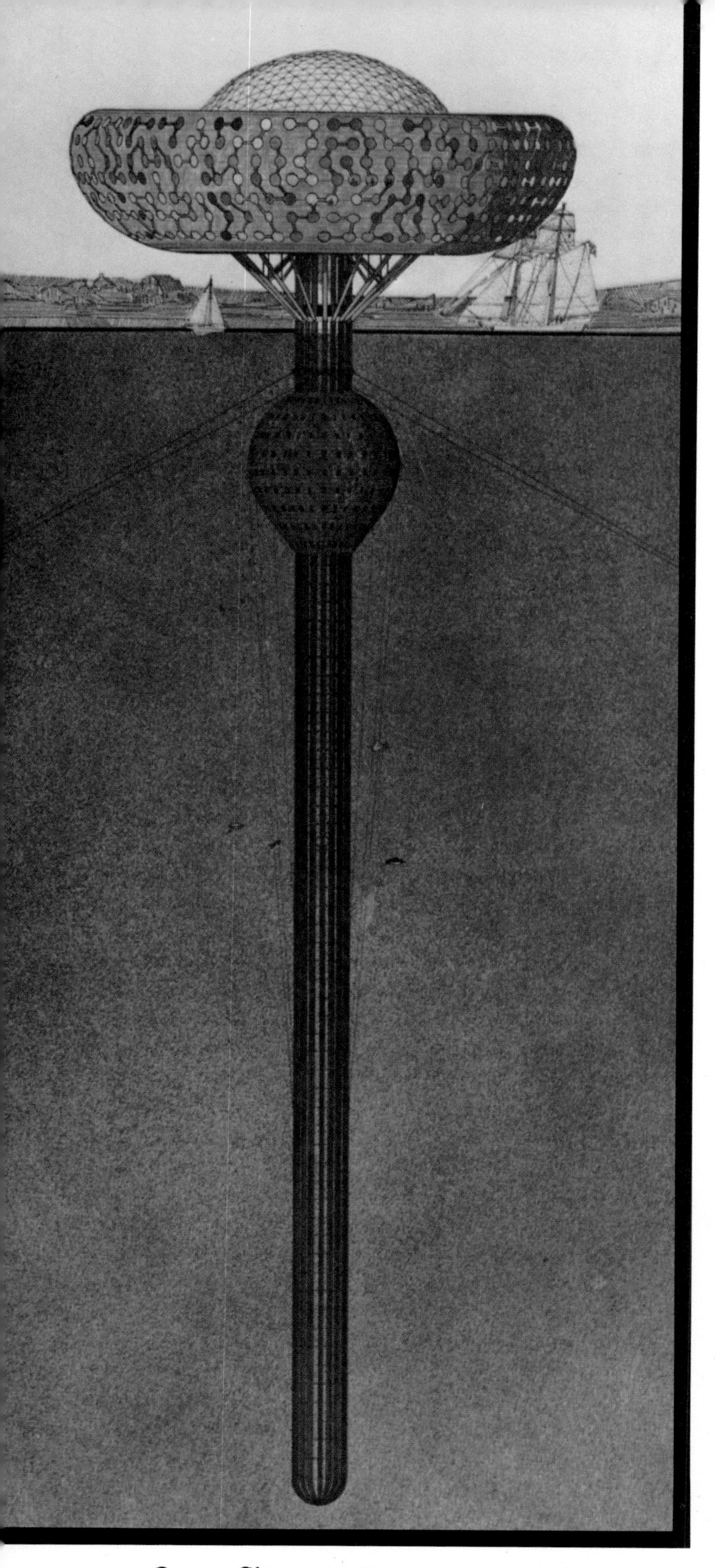

Ocean City, *a toplike structure containing apartments, a hotel, hospital, and school, was designed by Daniel Audrerie for Living Sea Corporation.*

Building on Water

Some of the most valuable and sought-after land in the world is that bordering on water, but there simply isn't enough waterfrontage to fill all the demands. In order to produce artificial offshore property, two families of projects have been proposed: artificial islands constructed on piles, and enormous floating structures.

Sea City, proposed by British architects and engineers, would be an offshore island built on piles of concrete. It would be capable of housing 30,000 people. A 16-story amphitheater would surround a large lagoon with clusters of islands. The city would provide all of the amenities of a seaside resort. Plans are well advanced for a number of other offshore structures. They will be built to play diverse roles: resort hotels, deepwater ports, airports, nuclear energy plants, or oceanographic observation laboratories.

For the operations that nobody wants as his neighbor—because of noise, the possibilities of oil spills, or any of countless other hazards—the artificial island may be a solution. However, it should be remembered that all the problems that make these operations so unloved on land will not disappear when they are moved out to sea.

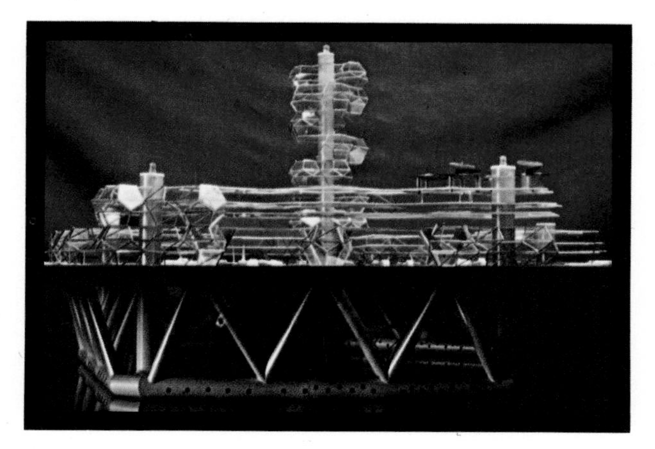

A floating island named ***Acrocean*** *would be stable in all sea conditions. The plan was made by French architect Edouard Albert.*

Energy from the Strait

Many years ago it was proposed that a dam be built across the Strait of Gibraltar. The water of the Mediterranean would evaporate in a short time, and when the level was 100 feet below that of the Atlantic, water would again be allowed to flow in. This flow could be harnessed to produce electric power. The idea was full of flaws. For one thing, it would kill most of the Mediterranean's marine life. It would close that sea for shipping, and its ports and seaside cities would be separated from the water by miles of stinking beaches covered with rotting marine life.

There is a better way we could take advantage of the strong currents that flow through the strait. It would require no technological breakthrough, but huge investments to install low pressure turbines, an aquatic replica of wind mills. At Rance River in northern France, low pressure turbines are run by the ebb and flow of tidal water; the plant has proved successful and will be expanded.

Water flowing through the strait at the surface moves in an easterly direction toward the hot, arid lands of the Middle East. Desert winds and hot sun evaporate the seawater, making the remaining water more saline. This saltier, heavier water sinks and sets up a countercurrent that moves west. The flow of water is reliable and strong. Hundreds of turbines could be placed in both currents in the strait to produce fuel-free, nonpolluting electric power. It would not interfere with shipping, since it could be installed below the draft depth of the largest ships. The topography at Gibraltar is compatible with such a complex generating plant. Similar projects are suggested for the English Channel and the Straits of Florida.

*The **Rock of Gibraltar** stands guard over a potential source of clean energy—powerful currents.*

Index

ILLUSTRATIONS AND CHARTS:

Howard Koslow—64, 65, 134, 135, 138.

PHOTO CREDITS:

Gail Ash—140-141; The Bettmann Archive, Inc.—74, 75, 78, 81, 118, 119, 128-129 (bottom); Bruce Coleman Inc.: Jen and Des Bartlett—17, Alan Blank—12-13 (top), Jane Burton—42, 45, Jeff Foott—43, David Hughes—11, 14, 15, R. M. Mariscal—46, Oxford Scientific Films—25 (top), Allan Power—56, W. Stoy—41, M. Tonooka—127, D. and K. Urry—16 (bottom); Czartoryska Photos: N. C. Flemming—70, 71; H. Edgerton—106-107, 111; Nat Fain—20 (top); Freelance Photographers Guild: Duane D. Davis—12 (bottom), Bob Gladden—32 (bottom), 39, Franz Lazi—98-99; Giraudon—76, 116; Thor Heyerdahl—96, 97; Marcel Ichac—110; International Hydrodynamics Company—87 (bottom); Paul Kanciruk and William Hernkind—18-19; Robert Marx—66-67; Richard C. Murphy—23, 24 (bottom), 28-29, 36-37, 47 (bottom), 130; NASA—136; National Maritime Museum, Greenwich—83; Naval Photographic Center—84, 85, 88 (top), 91 (top right), 120, 121; *The New York Times*—117; Nancy Palmer Photo Agency: Peter Throckmorton—108, 109; Photo communiquée par le Gouvernement de la République Populaire d' Albanie—70; Photo Ellebe—131; Guy C. Powell—54; The Sea Library: California Department of Fish and Game—51 (bottom), 58, B. Campoli—91 (top left), Jim and Cathy Church—34 (top), 35 (bottom), 44 (bottom), 125, Pat Colins, Institute of Marine Science, Miami, Fla.—16 (top), Jim Cooluris—2-3, Ben Cropp—53, Jack Drafahl—30-31 (top left), 30 (bottom), 44 (top), 61, Robert Evans—50 (top insert), 51 (top), 137, George Green—24 (top), Bill MacDonald—124 (bottom left), Tom McHugh, Marineland of Florida—60, Jack McKenney—115, 142, Robert Marx—104-105 (top), 104 (bottom right), Jim Morin—50 (top), Naval Undersea Center—86, 89 (top), 91 (bottom), Elliott Norse—20 (bottom), Carl Roessler—26-27, 33 (middle left), 33 (right), 57, S. Shane—104 (middle right), 105 (middle left), Paul Tzimoulis—34 (bottom), 105 (top right), 122-123 (bottom), 123 (top), 124 (bottom right); Eugene A. Shinn—21, 31 (right), 33 (bottom left); Siebe, Gorman & Company Ltd. (from *Deep Diving and Submarine Operations* by Sir Robert H. Davis)—82; Tom Stack & Associates: Ron Church—90, Ben Cropp—55, Dave La Touche—25 (bottom), Fred Livingston—73, Tom Myers—13 (bottom), 47 (top), Brian O'Heir—5, Tom Stack— 63, Ron Taylor—133; Taurus Photos: Eugene A. Shinn—129 (top), Dave Woodward—31 (bottom), 35 (top), 105 (bottom); Paul Tzimoulis—33 (top left), 59; UNESCO—68, 69; U.S. Navy Photograph—88 (bottom), 89 (bottom); Myron Wang—32 (top); Wide World Photos—80, 87 (top), 100; © D. P. Wilson—48, 49; World Life Research Institute: Devon Ludwik—38.